SO-BFD-448

Test Yourself

Precalculus Mathematics

Mark Weinfeld, M.S.
MathWorks
Union, NJ

Contributing Editors

Thomas A. Brown, Ph.D.
Phoenix, AZ

Tony Julianelle, Ph.D.
University of Vermont
Burlington, VT

Christi Heuer
Newtown, CT

NTC LEARNINGWORKS
NTC/Contemporary Publishing Group

Library of Congress Cataloging-in-Publication Data

Weinfeld, Mark.
    Precalculus mathematics / Mark Weinfeld ; contributing editors,
Thomas A. Brown, Tony Julianelle, and Christi Heuer. — Rev. ed.
        p.   cm. — (Test yourself)
    ISBN 0-8442-2382-4
    1. Mathematics.   I. Brown, Thomas A., Ph.D.   II. Julianelle,
Tony.   III. Heuer, Christi.   IV. Title.   V. Series: Test yourself
(Lincolnwood, Ill.)
    QA39.2.W412   1999
    510—dc21   98-54447
                                                        CIP

A *Test Yourself Books, Inc.* Project

Published by NTC LearningWorks
A division of NTC/Contemporary Publishing Group, Inc.
4255 West Touhy Avenue, Lincolnwood (Chicago), Illinois 60646-1975 U.S.A.
Copyright © 2000 by NTC/Contemporary Publishing Group, Inc.
Printed in the United States of America
International Standard Book Number: 0-8442-2382-4
    18 17 16 15 14 13 12 11 10 9 8 7 6 5 4 3 2 1

Contents

Preface

Test Yourself in Precalculus Mathematics provides a source of practice problems for all mathematics topics typically covered in a precalculus course. Whether you are in school and enrolled in a formal course or seeking to brush up on precalculus material on your own, you will find this book extremely helpful.

If you are currently studying precalculus mathematics in class, this book will serve as a very useful supplement to your course textbook. In order to master any subject area in mathematics, you must practice as much as you can. Use the book to test your comprehension of topics just completed in your course or as a way of reviewing for an upcoming test. In addition, there may be a particular type of problem with which you are struggling; you will probably find many problems of that type to work on in this book.

If your goal is to brush up on precalculus mathematics on your own, you will find this book useful as well. After working on the problems and checking your answers, the book will help you determine if it is time to move on or if you need to restudy certain problem types.

Whether you are in a course or studying on your own, be sure to study the problem solutions carefully—even those solutions for the problems you answered correctly. The solutions will help you understand the best way to think about similar problems when you see them in the future.

This book is for Janet, for all sorts of reasons.

Mark Weinfeld

How to Use This Book

This "Test Yourself" book is part of a unique series designed to help you improve your test scores on almost any type of examination you will face. Too often, you will study for a test—quiz, midterm, or final—and come away with a score that is lower than anticipated. Why? Because there is no way for you to really know how much you understand a topic until you've taken a test. The *purpose* of the test, after all, is to test your complete understanding of the material.

The "Test Yourself" series offers you a way to improve your scores and to actually test your knowledge at the time you use this book. Consider each chapter a diagnostic pretest in a specific topic. Answer the questions, check your answers, and then give yourself a grade. Then, and only then, will you know where your strengths and, more important, weaknesses are. Once these areas are identified, you can strategically focus your study on those topics that need additional work.

Each book in this series presents a specific subject in an organized manner, and although each "Test Yourself" chapter may not correspond to exactly the same chapter in your textbook, you should have little difficulty in locating the specific topic you are studying. Written by educators in the field, each book is designed to correspond, as much as possible, to the leading textbooks. This means that you can feel confident in using this book and that regardless of your textbook, professor, or school, you will be much better prepared for anything you will encounter on your test.

Each chapter has four parts:

Brief Yourself. All chapters contain a brief overview of the topic that is intended to give you a more thorough understanding of the material with which you need to be familiar. Sometimes this information is presented at the beginning of the chapter, and sometimes it flows throughout the chapter, to review your understanding of various *units* within the chapter.

Test Yourself. Each chapter covers a specific topic corresponding to one that you will find in your textbook. Answer the questions, either on a separate page or directly in the book, if there is room.

Check Yourself. Check your answers. Every question is fully answered and explained. These answers will be the key to your increased understanding. If you answered a question incorrectly, read the explanations to *learn* and *understand* the material. You will note that at the end of every answer you will be referred to a specific subtopic within that chapter, so you can focus your studying and prepare more efficiently.

Grade Yourself. At the end of each chapter is a self-diagnostic key. By indicating on this form the numbers of those questions you answered incorrectly, you will have a clear picture of your weak areas.

There are no secrets to test success. Only good preparation can guarantee higher grades. By utilizing this "Test Yourself" book, you will have a better chance of improving your scores and understanding the subject more fully.

Fundamentals

1

Brief Yourself

The Real Number System

Real numbers are used throughout precalculus. Within the real number system, groups of numbers of different types are identified and named.

The numbers that are used for counting

$$1, 2, 3, 4, 5, 6, \ldots$$

are called *natural numbers*. The set of natural numbers is symbolized by the letter *N*. The natural numbers, together with 0 and

$$-1, -2, -3, -4, -5, -6, \ldots$$

are called *integers*, symbolized by the letter *Z*. A real number is called a *rational number* if it can be expressed as the ratio of two integers. Thus,

$$\frac{1}{2}, 12, -45, \frac{-3}{2}, 0$$

are examples of rational numbers. The set of *irrational numbers* contains the real numbers that cannot be expressed as the ratio of two integers. Numbers such as

$$\sqrt{2}, \ -\pi, \ -\sqrt{7}$$

are irrational. The rational numbers together with the irrational numbers make up the set of *real numbers*.

The Order of Operations

When numerical expressions contain more than one operation, a procedure has been established to specify the order in which the operations should be performed. This procedure, called the *order of operations*, is given on the following page.

1. Perform all operations within parentheses or within other grouping symbols such as radicals and fraction bars.

2. Evaluate all exponents.

3. Perform all multipications and divisions in the order they appear from left to right.

4. Perform all additions and subtractions in the order they appear from left to right.

Integral Exponents

If n is a positive integer, the notation x^n is defined by

$$x^n = x\,x\,x\,x\ \dots\ x \ (n \text{ factors of } x)$$

We further define $x^0 = 1$ and $x^{-n} = \dfrac{1}{x^n}$.

There are five laws of exponents:

1. $x^a \times x^b = x^{a+b}$

2. $(x^a/x^b) = x^{a-b} \ (x \neq 0)$

3. $(x^a)^b = x^{ab}$

4. $(xy)^a = x^a y^a$

5. $(x/y)^a = x^a/y^a$

Special Products and Factors of Polynomials

An *algebraic term* is the product of numbers and variables, such as $5x$, $-2x^3$, or $17x^2y^5$. An expression containing two terms connected by addition or subtraction is called a *binomial*. An expression containing three terms connected by addition or subtraction is called a *trinomial*. Any expression containing more than one term is called a *polynomial*.

To multiply a binomial by a single term, use the distributive property:

$$A(B + C) = AB + AC$$

Two binomials can be multiplied using the following formula:

$$(A + B)(C + D) = AC + AD + BC + BD$$

There are several products that are frequently needed and should be memorized:

$$(a+b)^2 = a^2 + 2ab + b^2$$

$$(a - b)^2 = a^2 - 2ab + b^2$$

$$(a + b)^3 = a^3 + 3a^2b + 3ab^2 + b^3$$

$$(a - b)^3 = a^3 - 3a^2b + 3ab^2 + b^3$$

should be:
$a^3 - 3a^2b + 3ab^2 - b^3$

$$(a - b)(a + b) = a^2 - b^2$$

Factoring a polynomial refers to rewriting the polynomial as a product of factors. This can be accomplished by multiplying "in reverse." For example, since

$$5q(x + y^3) = 5qx + 5qy^3$$

if we are given $5qx + 5qy^3$, we can factor it as $5q(x + y^3)$. Similarly, the trinomial $a^2 - 2ab + b^2$ can be factored as $(a - b)^2$.

Roots and Radicals

If for any real number r we have $r^2 = x$, then we write $r = \sqrt{x}$ and say that r is the square root of x. Higher roots can be defined similarly. For example, if for any real number r we have $r^3 = x$, then we write $r = \sqrt[3]{x}$ and say that r is the cube root of x.

Note that if n is even

$\sqrt[n]{a}$ is real when $a \geq 0$
$\sqrt[n]{a}$ is not real when $a < 0$

If n is odd

$\sqrt[n]{a}$ is alway real and is defined such that $\left(\sqrt[n]{a} \right)^n = a$
$\equiv \sqrt[n]{p^n} = |p|$ if n is even
p if n is odd

Rational Exponents

We previously defined integral exponents. The definition of exponents can be extended to include rational exponents by means of the following definitions:

If n is a natural number

$$x^{1/n} = \sqrt[n]{x} \text{ when } x \geq 0 \text{ if } n \text{ is even}$$
$$x^{m/n} = (\sqrt[n]{x})^m \text{ or } x^{m/n} = \sqrt[n]{x^m}$$

Complex Numbers

The imaginary number system is useful in more advanced areas of mathematics. It is based on the definition of i, called the imaginary unit, as $i = \sqrt{-1}$. Thus, $i^2 = -1$, $i^3 = i \cdot i^2 = -i$, and $i^4 = i^2 \cdot i^2 = (-1)(-1) = 1$. Further, since $i^5 = (i^4)(i)$, we can see that $i^5 = (1)(i) = i$, and so on.

An *imaginary number* is any number of the form bi, where b is a real number and i is the imaginary unit. A *complex number* is any number of the form $a + bi$, where a and b are real numbers and i is the imaginary unit.

Test Yourself

Choose the best answer for each problem below. Simplify the expressions in problems 1–20.

1. $\sqrt{15 + 12 - 2}$

   A. 27

   B. 25

   C. 9

   D. 5

2. $\sqrt{16} + \sqrt{9} + \sqrt{1}$

   A. 5

   B. 8

   C. $\sqrt{26}$

   D. 10

3. $\left(\sqrt{3.8^2}\right)$

   A. 3.8

   B. 14.44

   C. 1.95

   D. 2.06

4. $(-4)^2 + \dfrac{16}{4} - 3 - (-4)^2$

   A. 6

   B. 5

   C. −3

   D. 1

5. -6^3

   A. 216

   B. −216

   C. $\dfrac{1}{216}$

   D. $\dfrac{-1}{216}$

6. $\dfrac{1}{-6^3}$

   A. 216

   B. −216

   C. $\dfrac{1}{216}$

   D. $\dfrac{-1}{216}$

7. $\dfrac{1}{4^2 + 2^3}$

   A. $\dfrac{1}{6}$

   B. $\dfrac{1}{14}$

   C. $\dfrac{1}{20}$

   D. $\dfrac{1}{24}$

8. $\dfrac{1}{(6+3)^3}$

   A. $\dfrac{1}{729}$

   B. $\dfrac{1}{243}$

   C. $\dfrac{1}{249}$

   D. $\dfrac{1}{432}$

9. $4^{-3} + 4^3$

   A. 1

   B. 0

   C. $64\dfrac{1}{64}$

   D. $\dfrac{1}{64}$

10. $\left| \dfrac{1}{7} \right|$

    A. $\dfrac{1}{7}$

    B. $\dfrac{-1}{7}$

    C. $\dfrac{1}{-7}$

    D. 7

11. $\left| \dfrac{-1}{7} \right|$

    A. $\dfrac{1}{7}$

    B. $\dfrac{-1}{7}$

    C. 7

    D. −7

12. $-\left| \dfrac{1}{7} \right|$

    A. $\dfrac{1}{7}$

    B. $\dfrac{-1}{7}$

    C. 7

    D. −7

13. $\left|\left(\dfrac{-1}{5}\right)^2\right|$

    A. $\dfrac{-1}{25}$

    B. $\dfrac{1}{25}$

    C. $\dfrac{1}{10}$

    D. $\dfrac{-1}{5}$

14. $\left(\dfrac{-1}{5}\right)^3$

    A. $\dfrac{1}{15}$

    B. $\dfrac{-1}{15}$

    C. $\dfrac{1}{125}$

    D. $\dfrac{-1}{125}$

15. $\dfrac{1}{2} \div \left(\dfrac{2}{7} \div \dfrac{4}{5}\right)$

    A. $\dfrac{5}{7}$

    B. $\dfrac{8}{7}$

    C. $\dfrac{7}{8}$

    D. $\dfrac{7}{5}$

16. $\left(\dfrac{1}{2} \div \dfrac{2}{7}\right) \div \dfrac{4}{5}$

    A. $\dfrac{4}{35}$

    B. $\dfrac{35}{4}$

    C. $\dfrac{7}{4}$

    D. $\dfrac{35}{16}$

17. $\left(\dfrac{1}{2} - \dfrac{2}{3}\right)^3$

    A. $\dfrac{1}{216}$

    B. $\dfrac{-1}{216}$

    C. $\dfrac{-1}{18}$

    D. $\dfrac{1}{6}$

18. $\dfrac{\frac{3}{4}}{3}$

    A. 4

    B. $\dfrac{1}{3}$

    C. $2\dfrac{1}{4}$

    D. 3

19. $(1.5)\left(\dfrac{3}{7}\right)$

    A. $\dfrac{14}{9}$

    B. 0.7

    C. $\dfrac{9}{14}$

    D. 2.1

20. $(3.2)\left(\dfrac{5}{4}\right)\left(\dfrac{1}{9}\right)(-2.5)$

    A. $\dfrac{167}{234}$

    B. $\dfrac{-5}{9}$

    C. 112

    D. $\dfrac{-10}{9}$

21. Expand $(3x + 2y)^2$.

    A. $6x + 4y$

    B. $9x^2 + 4y^2$

    C. $9x^2 + 12xy + 4y^2$

    D. $9x^2 + 5xy + 4y^2$

    E. $3x + 6xy + 4x^2$

22. Use the distributive law to rewrite $3x(4y + 6z)$.

    A. $3x + 4y + 6z$

    B. $12xy + 18xz$

    C. $7xy + 9xz$

    D. $12xy + 9xz$

    E. $12y^4 + 18x^7$

23. Use the associative law to rewrite $(9x)y^2$.

    A. $9xy^2$

    B. $9(xy)^2$

    C. $81xy^2$

    D. $(9xy)^2$

Rewrite the expressions in problems 24–32 so that all the negative exponents are eliminated:

24. $x^{-3}y^5x^3$

    A. x^4y^5

    B. y^5

    C. x^6y^5

    D. x^9y^5

    E. x^5y^5

25. $(x^2y^3)^{-2}$

    A. $\dfrac{1}{(x^4y^6)}$

    B. x^4y^6

    C. $\dfrac{1}{(x^2y^3)^2}$

    D. $\dfrac{-2}{(x^2y^3)}$

    E. $\dfrac{(x^2y^3)}{-2}$

26. $\left(\dfrac{x^{-2}y^3}{z^{-5}}\right)^{-4}$

    A. $\dfrac{z^{20}}{x^8y^{12}}$

    B. $\dfrac{z^{20}y^{12}}{x^8}$

    C. $\dfrac{x^8y^{12}}{z^{20}}$

    D. $\dfrac{x^8z^{20}}{y^{12}}$

    E. $\dfrac{x^8}{y^{12}z^{20}}$

27. $(x^4 + y^6)^{-2}$

    A. $\dfrac{2}{x^8 + y^{12}}$

    B. $x^{12} + y^{12}$

    C. $\dfrac{1}{(x^8 + 2x^4y^6 + y^{12})}$

    D. $x^{-8}y^{-12}$

28. $(3^2xyz^2)^{-3}$

    A. $\dfrac{1}{3x^3y^3z^6}$

    B. $\dfrac{1}{3^6x^3y^3z^6}$

    C. $\dfrac{1}{3^2xyz^2 - 3}$

    D. $3^6x^3y^3z^6$

29. $\dfrac{4^{-2} + 4^{-1} + 4^{-3}}{4^{-3}}$

    A. 46

    B. 43

    C. $\dfrac{1}{4^3}$

    D. 21

30. $\left(x^{3/2}\right)^4$

    A. $\dfrac{1}{x^6}$

    B. $x^{11/2}$

    C. $\dfrac{1}{x^{11/2}}$

    D. x^6

    E. 0

31. $\dfrac{x^{1.5}x^{-3.2}}{x^{-2.6}x^{4.7}}$

    A. x^{12}

    B. $\dfrac{1}{x^{3.8}}$

    C. $\dfrac{1}{x^{2.6}}$

    D. $x^{2.6}$

32. $(x^{1.5}x^{-3.2})^{2.7}$

    A. $x^{-12.96}$

    B. $x^{4.59}$

    C. $x^{12.96}$

    D. $\dfrac{1}{x^{4.59}}$

    E. 1

Factor the expressions in problems 33–49.

33. $3x^2y^5 + 9x^4y^4 - 12xy^3$

    A. $3xy\,(x^2y^5 - 3x^3y^3 - 4xy^3)$

    B. $3x^2y^5(1 + 3xy - 4xy^2)$

    C. $3xy^3(xy^2 + 3x^3y - 4)$

    D. $xy(3xy^4 + 9x^3y^3 - 12xy^2)$

34. $6a^4b^5c^4 - 2abc + 7a^2b^2c^2$

    A. $a^2b^2c^2(6a^2b^3c^2 - 2abc + 7)$

    B. $a^2b^3c^2(6a^2b^3c^2 - abc + 7)$

    C. $2abc(3a^3b^4c^3 - abc + 3abc)$

    D. $abc\,(6a^3b^4c^3 - 2 + 7abc)$

35. $25x^2 - 25y^2$

    A. $25(x + y)(x - y)$

    B. $5(x + y)(x - y)$

    C. $25(x - y)^2$

    D. $25(x + y)^2$

36. $14x^2y2 - 4xy + 7x^2y$

    A. $xy(14xy - 4 + 7x)$

    B. $14(x^2y^2 - xy + x^2y)$

    C. $xy(2xy - 1 + x)$

    D. $7xy(2xy - 2 + x^2)$

37. $x^4 - x^3 + 2x - 2$

    A. $x(x^3 - x^2 + 2)$

    B. $(x^3 + 2)(x - 1)$

    C. $(x + 2)(x^3 - 1)$

    D. $(x + 2)^3$

38. $x^3 - x^2 - x + 1$

    A. $(x - 1)^2(x + 1)$

    B. $(x - 1)^3$

    C. $(x + 1)(x - 1)(x)$

    D. $(x + 1)^3$

39. $16x^4 - 9$

    A. $(2x - 3)^4$

    B. $(2x + 3)^2(2x - 3)^2$

    C. $(4x^2 - 3)(4x^2 + 3)$

    D. $(4x - 3)(4x^2 + 3)$

40. $x^2 - 2x + 1$

    A. $(x + 1)^2$

    B. $(x - 1)(x + 1)$

    C. $(x + 1)(x - 2)$

    D. $(x - 1)^2$

41. $x^2 - 5x + 6$

    A. $(x - 2)(x + 3)$

    B. $(x - 2)(x - 3)$

    C. $(x + 2)(x + 3)$

    D. $(x - 1)(x + 6)$

42. $x^2 - 1$

    A. $x(x - 1)$

    B. $x^2(x - 1)$

    C. $(x + 1)(x - 1)$

    D. $(x - 1)^2$

43. $6x^2 + x - 1$

    A. $(2x + 1)(3x + 1)$

    B. $(2x - 1)(3x + 1)$

    C. $(2x + 1)(3x - 1)$

    D. $(6x + 1)(x + 1)$

44. $6x^4 + x^3 - x^2$

    A. $x^2(2x + 1)(3x - 1)$

    B. $x(x + 1)(6x + 2)$

    C. $(2x^2 + 1)(3x^2 - 1)$

    D. $(6x + 1)(x^3 - 1)$

45. $6x^2 + 11x - 10$

    A. $(6x - 1)(x - 10)$

    B. $(2x - 5)(3x - 2)$

    C. $(2x + 5)(3x - 2)$

    D. $(6x + 1)(x - 10)$

46. $81x^8 - 16y^{12}$

    A. $(3x + 4y)(3x - 4y)^3$

    B. $(3x^2 - 2y^3)(3x^2 + 2y^3)(9x^4 + 4y^6)$

    C. $(9x^4 - 4y^6)(9x^4 + 4y^6)$

    D. $(9x^4 - 4y^6)^2$

47. $-y^4 + x^4$

    A. $(-y^2 + x^2)(-y^2 - x^2)$

    B. $(x + y)(x - y)(x^2 + y^2)$

    C. $(x - y)^4$

    D. $(y - x)^4$

48. $t^2z^4 - tz + 3t^2z^5$

    A. $tz(t z - 1 + 3tz^4)$

    B. $t^2z^4(1 - t^{-1}z^{-1} + 3tz)$

    C. $t^2z(z^{-1} + 3t^{-1}z^3)$

    D. $tz(tz^3 - 1 + 3tz^4)$

49. $t^4 - 1$

    A. $(t^2 + 1)(t - 1)(t + 1)$

    B. $(t + 1)(t^3 - t + 1)$

    C. $(t - 1)(t^3 - t + 1)$

    D. $(t^2 - 1)^2$

50. Subtract: $(3x^2 + 5x + 7) - (2x^2 + 5x + 7)$

    A. $5x^2 + 10x + 14$

    B. x^2

    C. 0

    D. $x^4 + 10x^2$

51. Add: $(-2x^3 + 7x^2 + 2x) + (4x^3 + x + 7)$

    A. $2x^3 - 7x^2 + 9x$

    B. $2x^6 + 8x^3 + 14x$

    C. $2x^3 + 7x^2 + 3x + 7$

    D. $-8x^6 + x^4 + 14x$

52. Multiply: $x^3(x^4 - x^3 + 7x)x^2$

    A. $x^9 - x^8 + 7x^6$

    B. $x^{14} - x^{11} + 21x^5$

    C. $x^7 - x^9 + 7x^3$

    D. $x^{20} - x^{15} + 7x^5$

53. Multiply: $(x^2 + 1)(x^3 + 2x^2 - x)x$

    A. $x^7 + 2x^6 + 2x^4 - x^3$

    B. $x^4 + 2x^3 - x^2$

    C. $x^5 + 2x^4 + 2x^2 - x$

    D. $x^6 + 2x^5 + 2x^3 - x^2$

54. Multiply: $(3x^2 - x^3)(2x - 7)$

    A. $6x^3 + 6x + 7$

    B. $2x^4 + 13x^3 - 21x^2$

    C. $x^4 + 12x - 7x^3$

    D. $-2x^4 + 13x^3 - 21x^2$

55. Expand: $(x - 1)^3$

    A. $x^3 - 3x^2 + 3x - 1$

    B. $x^3 - 1$

    C. $x^3 - 3x^2 - 3x - 1$

    D. $x^3 - 3x$

56. Expand: $(x - 1)^2(x + 1)$

    A. $x^3 + x^2 + 3x - 1$

    B. $x^3 - x^2 - x + 1$

    C. $x^3 + 3x - 1$

    D. $x^3 - 3x + 1$

57. Simplify: $7i - 3i$

    A. $-4i$

    B. 4

    C. $4i$

    D. -4

58. Simplify: $(-2i)^3$

    A. -8

    B. $-8i$

    C. 8

    D. $8i$

59. Simplify: $4i^3 + 7i^3$

    A. $-11i$

    B. $11i$

    C. 11

    D. -11

60. Simplify: $(6i^2 - 4i^4)^2$

    A. 100

    B. $2i^8$

    C. $-2i^8$

    D. $2i^6$

61. Simplify: $(3i^{16})^4$

    A. $81i^{20}$

    B. 81

    C. -81

    D. $27i$

62. Simplify: i^{121}

    A. $-i$

    B. i

    C. 1

    D. -1

63. Simplify: $i^{640} - 1$

    A. i

    B. -1

    C. 0

    D. 1

64. Simplify: $4i^8 - 2i^4 - 2$

    A. i^2

    B. 2

    C. 1

    D. 0

65. Simplify: $\sqrt{-9}$

    A. -3

    B. $3i$

    C. $-3i$

    D. 3

66. Simplify: $\sqrt{-27}$

    A. $-3\sqrt{3}$

    B. $3\sqrt{3}$

    C. $3i\sqrt{3}$

    D. $-3i\sqrt{3}$

67. Factor: $v^{2/3} - 1$

    A. $(v^3 - 1)(v^{1/3} + 1)$

    B. $(v^{1/2} - 1)(v^{2/3} + 1)$

    C. $(v^{1/3} - 1)^2$

    D. $(v^{1/3} - 1)(v^{1/3} + 1)$

68. Factor: $v^{4/5} - 9y^{2/3}$

    A. $(v^{1/3} - 3y^{1/3})^2$

    B. $(v^{2/5} - 3y^{1/3})(v^{2/5} + 3y^{1/3})$

    C. $(v^{1/5} - 3y^{1/3})(v^2 - 3y^2)$

    D. $(v^{2/5} - 6y^{1/3})(v^{2/5} + 3y^{1/3})$

69. Factor: $6x^2 - 7x - 20$

    A. $(6x + 1)(x - 20)$

    B. $(3x + 4)(2x - 5)$

    C. $(6x - 1)(x + 20)$

    D. $(3x - 4)(x - 20)$

70. Factor: $8t^3 - 27$

    A. $(2t + 3)(4t^2 + 6t + 9)$

    B. $(4t + 3)(2t^2 - 3t + 9)$

    C. $(2t - 3)(4t^2 + 6t + 9)$

    D. $(4t - 3)(2t^2 + 9)$

 # Check Yourself

1. D. $\sqrt{15+12-2} = \sqrt{25} = 5$ (**Square roots**)

2. B. $\sqrt{16} + \sqrt{9} + \sqrt{1} = 4 + 3 + 1 = 8$ (**Square roots**)

3. A. $\left(\sqrt{3.8^2}\right)^2 = 3.8$ (**Square roots**)

4. D. $(-4)^2 + \dfrac{16}{4} - 3 - (-4)^2 = 16 + 4 - 3 - 16 = 1$ (**Exponents**)

5. B. $-6^3 = -216$ (**Exponents**)

6. D. $\dfrac{1}{-6^3} = \dfrac{-1}{216}$ (**Exponents**)

7. D. $\dfrac{1}{4^2 + 2^3} = \dfrac{1}{16 + 8} = \dfrac{1}{24}$ (**Exponents**)

8. A. $\dfrac{1}{(6+3)^3} = \dfrac{1}{9^3} = \dfrac{1}{729}$ (**Exponents**)

9. C. $4^{-3} + 4^3 = \dfrac{1}{4^3} + 4^3 = \dfrac{1}{64} + 64 = 64\dfrac{1}{64}$ (**Exponents**)

10. A. $\left|\dfrac{1}{7}\right| = \dfrac{1}{7}$ (**Absolute value**)

11. A. $\left|\dfrac{-1}{7}\right| = \dfrac{1}{7}$ (**Absolute value**)

12. B. $-\left|\dfrac{1}{7}\right| = -\dfrac{1}{7}$ (**Absolute value**)

13. B. $\left|\left(-\dfrac{1}{5}\right)^2\right| = \left|\dfrac{1}{25}\right| = \dfrac{1}{25}$ (**Absolute value**)

14. D. $\left(-\dfrac{1}{5}\right)^3 = \dfrac{-1}{125}$ (**Multiplication with real numbers**)

15. D. $\dfrac{1}{2} \div \left(\dfrac{2}{7} \div \dfrac{4}{5}\right) = \dfrac{1}{2} \div \dfrac{5}{14} = \left(\dfrac{7}{5}\right)$ (**Multiplication with real numbers**)

16. D. $\left(\dfrac{1}{2} \div \dfrac{2}{7}\right) \div \dfrac{4}{5} = \dfrac{7}{4} \div \dfrac{4}{5} = \dfrac{7}{4} \cdot \dfrac{5}{4} = \dfrac{35}{16}$ (**Multiplication with real numbers**)

17. B. $\left(\dfrac{1}{2} - \dfrac{2}{3}\right)^3 = \left(\dfrac{3}{6} - \dfrac{4}{6}\right)^3 = \left(\dfrac{-1}{6}\right)^3 = \dfrac{-1}{216}$ (**Multiplication with real numbers**)

18. C. $\dfrac{\frac{3}{4}}{\frac{4}{3}} = \dfrac{\frac{3}{1}\left(\frac{3}{4}\right)}{\frac{4}{3}\left(\frac{3}{4}\right)} = \dfrac{9}{4} = 2\dfrac{1}{4}$ (**Multiplication with real numbers**)

19. C. $(1.5)\left(\dfrac{3}{7}\right) = \dfrac{3}{2}\left(\dfrac{3}{7}\right) = \dfrac{9}{14}$ (**Multiplication with real numbers**)

20. D. $(3.2)\left(\dfrac{5}{4}\right)\left(\dfrac{1}{9}\right)(-2.5) = \left(\dfrac{32}{4}\right)\left(\dfrac{1}{9}\right)\left(-\dfrac{5}{4}\right) = \dfrac{-10}{9}$ (**Multiplication with real numbers**)

21. C. $(3x + 2y)^2 = (3x + 2y)(3x + 2y) = (3x)(3x) + 6xy + 6xy + (2y)(2y) = 9x^2 + 12xy + 4y^2$ (**Multiplication with real numbers**)

22. B. $3x(4y + 6z) = (3x)(4y) + (3x)(6z) = 12xy + 18xz$ (**Basic laws of arithmetic**)

23. A. $(9x)y^2 = 9xy^2$ (**Basic laws of arithmetic**)

24. B. $x^{-3}y^5x^3 = x^{-3+3}y^5 = x^0y^5 = y^5$ (**Exponents**)

25. A. $(x^2y^3)^{-2} = x^{(2)(-2)}y^{(3)(-2)} = x^{-4}y^{-6} = \dfrac{1}{x^4y^6}$ (**Exponents**)

26. E. $\left(\dfrac{x^{-2}y^3}{z^{-5}}\right)^{-4} = \left(\dfrac{x^{(-2)(-4)}y^{(3)(-4)}}{z^{(-5)(-4)}}\right) = \dfrac{x^8y^{-12}}{z^{20}} = \dfrac{x^8}{y^{12}z^{20}}$ (**Exponents**)

27. C. $(x^4 + y^6)^{-2} = \dfrac{1}{(x^4 + y^6)^2} = \dfrac{1}{(x^8 + 2x^4 + y^6 + y^{12})}$ (**Exponents**)

28. B. $(3^2 xyz^2)^{-3} = 3^{(2)(-3)} x^{-3} y^{-3} z^{(2)(-3)} = 3^{-6} x^{-3} y^{-3} z^{-6} = \dfrac{1}{3^6 x^3 y^3 z^6}$ (**Exponents**)

29. D. $\dfrac{4^{-2} + 4^{-1} + 4^{-3}}{4^{-3}} = \dfrac{\dfrac{1}{4^2} + \dfrac{1}{4^1} + \dfrac{1}{4^3}}{\dfrac{1}{4^3}} = \dfrac{\dfrac{1}{16} + \dfrac{1}{4} + \dfrac{1}{64}}{\dfrac{1}{64}} = \dfrac{\dfrac{21}{64}}{\dfrac{1}{64}} = 21$ (**Exponents**)

30. D. $(x^{3/2})^4 = x^{(3/2)(4)} = x^6$ (**Exponents**)

31. B. $\dfrac{x^{1.5} x^{-3.2}}{x^{-2.6} x^{4.7}} = x^{1.5 - 3.2 + 2.6 - 4.7} = x^{-3.8} = \dfrac{1}{x^{3.8}}$ (**Exponents**)

32. D. $(x^{1.5} x^{-3.2})^{2.7} = (x^{1.5 - 3.2})^{2.7} = (x^{-1.7})^{2.7} = x^{(-1.7)(2.7)} = x^{-4.59} = \dfrac{1}{x^{4.59}}$ (**Exponents**)

33. C. $3x^2 y^5 + 9x^4 y^4 - 12xy^3 = 3xy^3 (xy^2 + 3x^3 y - 4)$ (**Factoring**)

34. D. $6a^4 b^5 c^4 - 2abc + 7a^2 b^2 c^2 = abc\,(6a^3 b^4 c^4 - 2 + 7abc)$ (**Factoring**)

35. A. $25x^2 - 25y^2 = 25(x^2 - y^2) = 25(x+y)(x-y)$ (**Factoring**)

36. A. $14x^2 y^2 - 4xy + 7x^2 y = xy(14xy - 4 + 7x)$ (**Factoring**)

37. B. $x^4 - x^3 + 2x - 2 = x^3(x-1) + 2(x-1) = (x^3 + 2)(x-1)$ (**Factoring**)

38. A. $x^3 - x^2 - x + 1 = x^2(x-1) - 1(x-1) = (x^2 - 1)(x-1) = (x-1)(x+1)(x-1) = (x-1)^2(x+1)$ (**Factoring**)

39. C. $16x^4 - 9 = (4x^2 - 3)(4x^2 + 3)$ (**Factoring**)

40. D. $x^2 - 2x + 1 = (x-1)^2$ (**Factoring**)

41. B. $x^2 - 5x + 6 = (x-2)(x-3)$ (**Factoring**)

42. C. $x^2 - 1 = (x+1)(x-1)$ (**Factoring**)

43. C. $6x^2 + x - 1 = (2x+1)(3x-1)$ (**Factoring**)

44. A. $6x^4 + x^3 - x^2 = x^2(2x+1)(3x-1)$ (**Factoring**)

45. C. $6x^2 + 11x - 10 = (2x + 5)(3x - 2)$ (**Factoring**)

46. B. $81x^8 - 16y^{12} = (9x^4 - 4y^6)(9x^4 + 4y^6) = (3x^2 - 2y^3)(3x^2 + 2y^3)(9x^4 + 4y^6)$ (**Factoring**)

47. B. $-y^4 + x^4 = x^4 - y^4 = (x^2 - y^2)(x^2 + y^2) = (x+y)(x-y)(x^2 + y^2)$ (**Factoring**)

48. D. $t^2z^4 - tz + 3t^2z^5 = tz(tz^3 - 1 + 3tz^4)$ **(Factoring)**

49. A. $t^4 - 1 = (t^2 + 1)(t - 1)(t + 1)$ **(Factoring)**

50. B. $(3x^2 + 5x + 7) - (2x^2 + 5x + 7) = (3x^2 - 2x^2) + (5x - 5x) + (7 - 7) = x^2$ **(Arithmetic of polynomials)**

51. C. $(-2x^3 + 7x^2 + 2x) + (4x^3 + x + 7) = -2x^3 + 4x^3 + 7x^2 + 2x + x + 7 = 2x^3 + 7x^2 + 3x + 7$ **(Arithmetic of polynomials)**

52. A. $x^3(x^4 - x^3 + 7x)x^2 = (x^7 - x^6 + 7x^4)x^2 = x^9 - x^8 + 7x^6$ **(Multiplication of polynomials)**

53. D. $(x^2 + 1)(x^3 + 2x^2 - x)x = (x^5 + 2x^4 - x^3 + x^3 + 2x^2 - x)x = (x^5 + 2x^4 + 2x^2 - x)x = x^6 + 2x^5 + 2x^3 - x^2$ **(Multiplication of polynomials)**

54. D. $(3x^2 - x^3)(2x-7) = 6x^3 - 21x^2 - 2x^4 + 7x^3 = -2x^4 + 13x^3 - 21x^2$ **(Multiplication of polynomials)**

55. A. $(x - 1)^3 = [(x - 1)(x - 1)](x - 1) = (x^2 - 2x+1)(x - 1) = x^3 - x^2 - 2x^2 + 2x + x - 1 = x^3 - 3x^2 + 3x - 1$ **(Multiplication of polynomials)**

56. B. $(x - 1)^2(x + 1) = [(x - 1)(x - 1)](x + 1) = (x^2 - 2x + 1)(x + 1) = x^3 + x^2 - 2x^2 - 2x + x + 1 = x^3 - x^2 - x + 1$ **(Multiplication of polynomials)**

57. C. $7i - 3i = 4i$ **(Complex arithmetic)**

58. D. $(-2i)^3 = (-2)^3(i)^3 = -8(i)i(i) = -8(-1)(i) = 8i$ **(Complex arithmetic)**

59. A. $4i^3 + 7i^3 = (4 + 7)i^3 = 11i^3 = 11(i)^2i = 11(-1)i = -11i$ **(Complex arithmetic)**

60. A. $(6i^2 - 4i^4)^2 = (6i^2 - 4)^2 = (-6 - 4)^2 = (-10)^2 = 100$ **(Complex arithmetic)**

61. B. $(3i^{16})^4 = (3(1))^4 = 3^4 = 81$ **(Complex arithmetic)**

62. B. $i^{121} = (i^4)^{30}i = 1^{30}i = i$ **(Complex arithmetic)**

63. C. $i^{640} - 1 = (i^4)^{160} - 1 = 1^{160} - 1 = 0$ **(Complex arithmetic)**

64. D. $4i^8 - 2i^4 - 2 = 4(i^4)^2 - 2(1)^4 - 2 = 4(1)^2 - 2 - 2 = 4 - 2 - 2 = 0$ **(Complex arithmetic)**

65. B. $\sqrt{-9} = \sqrt{9}\,\sqrt{-1} = 3i$ **(Complex arithmetic)**

66. C. $\sqrt{-27} = \sqrt{27}\sqrt{-1} = 3i\sqrt{3}$ **(Complex arithmetic)**

67. D. $v^{2/3} - 1 = (v^{1/3} - 1)(v^{1/3} + 1)$ **(Factoring)**

68. B. $v^{4/5} - 9y^{2/3} = (v^{2/5} - 3y^{1/3})(v^{2/5} + 3y^{1/3})$ **(Factoring)**

69. B. $6x^2 - 7x - 20 = (3x + 4)(2x - 5)$ **(Factoring)**

70. C. $8t^3 - 27 = (2t - 3)(4t^2 + 6t + 9)$ **(Factoring)**

Grade Yourself

Circle the numbers of the questions you missed, then fill in the total incorrect for each topic. If you answered more than three questions incorrectly, you need to focus on that topic. (If a topic has fewer than three questions and you had at least one wrong, we suggest you study that topic also. Read your textbook or a review book, or ask your teacher for help.)

Subject: Fundamentals

Topic	Question Numbers	Number Incorrect
Square roots	1, 2, 3	
Exponents	4, 5, 6, 7, 8, 9, 24, 25, 26, 27, 28, 29, 30, 31, 32	
Absolute value	10, 11, 12, 13	
Multiplication with real numbers	14, 15, 16, 17, 18, 19, 20, 21	
Basic laws of arithmetic	22, 23	
Factoring	33, 34, 35, 36, 37, 38, 39, 40, 41, 42, 43, 44, 45, 46, 47, 48, 49, 67, 68, 69, 70	
Arithmetic of polynomials	50, 51	
Multiplication of polynomials	52, 53, 54, 55, 56	
Complex arithmetic	57, 58, 59, 60, 61, 62, 63, 64, 65, 66	

Linear Equations and Inequalities

2

Brief Yourself

An *equation* is a mathematical statement that two quantities are equal. If the equation contains only numbers, it can be classified as either true or false. For example, the equation $7 + 12 = 19$ is true; the equation $7 + 13 = 19$ is false. An equation that contains one or more variables is neither true nor false, but is called *conditional*. A conditional equation is an equation that is true for only certain values of the variables. These values are said to *satisfy* the equation and are called *solutions* or *roots* of the equation. As an example, the equation $5x - 2 = 3$ is only true for the value x = 1. Thus, we say that 1 is the solution of the equation. To *solve an equation* means to find all of the values of the variables involved that make both sides equal. Learning to solve equations is one of the most fundamental and critical algebraic skills, since the technique enables us to solve a wide variety of numerical problems that cannot be solved using the techniques of ordinary arithmetic.

Equations can be classified by the number of variables they contain and the powers to which these variables are raised. An equation that only contains one variable that is raised to the first power is called a *linear* (or *first degree*) equation in one unknown. A linear equation in two unknowns contains two variables, each of which is raised to the first power. Thus, $5x - 7 = 21$ is a linear equation in one unknown, and $5x + 7y = 32$ is a linear equation in two unknowns. An equation that contains one unknown that is raised to the second power (and also possibly to the first power) is called a *quadratic* equation in one unknown. An example of such an equation is $5x^2 + 3x - 6 = 12$. In this section, we will consider how to solve linear equations.

A linear equation in one unknown can occasionally be solved by trial and error. However, it is generally easier to go about solving the equation in an organized manner, using what is known as the method of *inverse operations*. The idea behind this method is to take a given equation and transform it to an *equivalent* equation (that is to say, an equation with the same solution) that has the variable by itself on one side of the equation and a number by itself on the other. In order to isolate the variable, we must perform inverse operations on the side of the equation containing the variable to undo the operations on that side. Remember that addition undoes subtraction and vice versa, and multiplication undoes division and vice versa. Also remember that to avoid changing the solution to an equation, any operation performed on one side of the equation must be performed on the other side as well.

For example, consider the equation $5x + 3 = 18$. In order to isolate the x on the left-hand side of the equation, we first must get rid of the 3. Since the 3 is added to the left-hand side, we subtract (that is, perform the inverse operation) on both sides. When we subtract 3 from both sides, we obtain the equivalent equation $5x = 15$. Next, we must eliminate the 5 from the left-hand side. Since the 5 multiplies the x, we divide both sides by 5, obtaining the solution $x = 3$. The solution obtained can be checked by substituting it back into the original equation and making sure that both sides are equal.

Linear equations in one unknown can become much more complex than the one given in the example above, but they can all be solved using the technique of inverse operations. Should a linear equation contain parentheses, eliminate the parentheses by using the distributive property. Should the variable appear on both sides of the equation, use the method to move both variables to the same side and then combine them. If the equation contains fractions, it is usually easiest to begin by multiplying both sides of the equation by the least common denominator; this will get rid of the fractions and make it easier to proceed.

Formulas are mathematical statements that typically contain more than one variable. Often it is useful to solve a formula for one of its variables. For example, the formula for the area of a rectangle is given by $A = L \times W$. If, however, we need to find the width of the rectangle, it is more convenient to solve the equation for W, obtaining $W = \dfrac{A}{L}$. To solve an equation for any particular variable, simply treat all of the other variables as if they were numbers, and use the technique described above.

An *inequality* is a mathematical statement in which two quantities are related to each other by one of the following symbols: $>$ (greater than), $<$ (less than), $\geq$ (greater than or equal to), or $\leq$ (less than or equal to). Linear inequalities in one variable, like linear equations in one variable, are solved by the method of inverse operations. However, when solving inequalities, there is one additional rule that must be observed: whenever you multiply or divide the inequality by a negative number, the direction of the inequality sign must be reversed. Thus, when multiplying or dividing by a negative number, $<$ becomes $>$, and $\leq$ becomes $\geq$.

While a linear equation in one unknown typically has a single solution, a linear inequality in one unknown typically has a range of solutions. For example, the inequality $3x > 12$ is true for all values of $x > 4$. The solution set of an inequality can be visualized on a number line. For example, the solution set $x > 4$ can be indicated as follows:

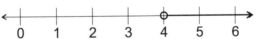

The open circle above the 4 indicates that 4 is not a solution to the inequality. In order to graph the inequality $x \geq 4$, we would make a closed circle on the 4.

In order to solve a linear equation in two unknowns, we must give pairs of values for the unknowns that when taken together satisfy the equation. For example, consider the equation $x + y = 10$. This equation is solved by the pair of values $x = 4$ and $y = 6$. Often, such a solution is written as an *ordered pair*, (4, 6), where, by convention, the first entry represents the value of x and the second entry represents the value of y. Unlike the linear equation in one unknown, a linear equation in two unknowns typically has an infinite number of solutions. Thus, while (4, 6) solves the equation, so does (5, 5), (6, 4), (10, 0), (.5, 9.5), etc. Usually the best technique to generate solutions for a linear equation in two unknowns is to solve the equations for one of the variables and then select values for the other variable. The corresponding values for the solved variable can be easily computed. In the next chapter, we will see how to graph the solution set of a linear equation in two unknowns and see that all of the solutions lie on a straight line.

Test Yourself

Classify the following equations as true, false, or conditional.

1. $3(7 + 2) = 27$

2. $17y - 21 = 4$

3. $12x + 7 - 2x = 10x + 7$

4. $3 + 2 = 21 \times \dfrac{2}{3}$

5. $3x - 23 = 17 + 5x$

Classify the following equations as linear or nonlinear.

6. $2x + 1 = 3x + 7$

7. $4x^2 + 3 = 2(6 - 3x + 2x^2)$

8. $2\sqrt{x} = 5 - 3\sqrt{x}$

9. $\dfrac{2x + 1}{4} = \dfrac{7}{3}$

10. $x(x + 7) = 3x + 5$

Solve the following linear equations for the variable indicated.

11. $x + 7 = 11$

12. $y - 9 = 5$

13. $12z = 108$

14. $\dfrac{x}{5} = 9$

15. $3x + 18 = 33$

16. $5y - 12 = 28$

17. $13 = \dfrac{z}{3} - 1$

18. $-1 = 11 - 2x$

19. $-8 = \dfrac{y}{4} + 7$

20. $6 = 32 + 4x$

Solve the following equations for the variable indicated.

21. $3(4x - 5) = -3$

22. $10 - 4(3x - 1) = 25$

23. $13 = 3 + 2(x - 7)$

24. $8y - 2y = 30$

25. $8x = -2x + 30$

26. $2(x - 5) + 3x = 15$

27. $17 - 2x = 9 - 3x$

28. $\dfrac{x}{2} + \dfrac{x}{3} = 25$

29. $\dfrac{y}{3} = \dfrac{y}{2} - 2$

30. $\dfrac{6}{x} + \dfrac{6}{4x} = \dfrac{5}{8}$

31. $\dfrac{2z}{5} = \dfrac{4}{3}$

32. $\dfrac{5x}{6} = \dfrac{4x - 3}{3}$

33. $\sqrt{6x + 2} + 6 = 10$

34. $\sqrt{2x} + 8 = 3$

35. $\sqrt{-3a} = \sqrt{-5a - 2}$

Translate each of the following statements into an equation.

36. Twice z plus 8 is equal to 18.

37. One-third of z is equal to three more than y.

38. Five more than twice w is equal to fourteen.

39. Six times a number decreased by four is equal to three times the number increased by two.

40. The product of 7 and x is one more than the quotient of x and 3.

Solve the following word problems.

41. If 12 times a number is decreased by 4, the result is the same as when 6 times the number is increased by 8. What is the number?

42. Three more than twice a certain number is equal to 49. Find the number.

43. The sum of three consecutive integers is 105. Find the smallest of these integers.

44. A piece of wood that is 36 inches long is cut into two pieces so that twice the longer piece is two inches more than three times the shorter piece. How long is the shorter piece?

45. Janet is now three times older than Brian. Four years ago, Janet was four times as old as Brian was then. How old is Janet now?

Solve the following formulas for the variables indicated.

46. $x = y - z$ for y

47. $d = 5e + 2f$ for e

48. $ax - b = 3c$ for x

49. $P = 2(L + W)$ for L

50. $\dfrac{q}{x} + \dfrac{p}{4x} = 2$ for x

Solve the following inequalities for the indicated variable and graph the solution on a number line.

51. $3x - 5 > 16$

52. $8x + 2 < 18$

53. $\dfrac{x}{6} - \dfrac{x}{2} \le 2$

54. $6y + 5 \ge 3(y - 2)$

55. Find five different solutions to the equation $2x + y = 6$.

56. If $x = 5$, find the corresponding value of y to solve the equation $5x - 2y = 7$.

57. Find three different solutions to the equation $2x - 3y = 6$.

58. If $y = 4$, find the corresponding value of x which will solve the equation $2x = 7y - 14$.

Check Yourself

1. $3(7 + 2) = 27$ is equivalent to $3(9) = 27$, which is true. (**Classifying equations**)

2. $17y - 21 = 4$ is conditional. (**Classifying equations**)

3. $12x + 7 - 2x = 10x + 7$ simplifies to $10x + 7 = 10x + 7$, which is true. (**Classifying equations**)

4. $3 + 2 = 21 \times \dfrac{2}{3}$ is equivalent to $5 = 14$, which is true. (**Classifying equations**)

5. $3x - 23 = 17 + 5x$ is conditional. (**Classifying equations**)

6. $2x + 1 = 3x + 7$ is linear. (**Classifying equations**)

7. $4x^2 + 3 = 2(6 - 3x + 2x^2)$ simplifies to $4x^2 + 3 = 12 - 6x + 4x^2$ or $3 = 12 - 6x$, which is linear. (**Classifying equations**)

8. $2\sqrt{x} = 5 - 3\sqrt{x}$ involves square roots of the variable x and thus is not linear. (**Classifying equations**)

9. $\dfrac{2x + 1}{4} = \dfrac{7}{3}$ contains the variable x to the first power and is thus linear. (**Classifying equations**)

10. $x(x + 7) = 3x + 5$ is equal to $x^2 + 7x = 3x + 5$, which is quadratic and, thus, not linear. (**Classifying equations**)

11. $x + 7 = 11$
    $x + 7 - 7 = 11 - 7$
    $x = 4$ (**Solving linear equations**)

12. $y - 9 = 5$
    $y - 9 + 9 = 5 + 9$
    $y = 14$ (**Solving linear equations**)

13. $12z = 108$
    $\dfrac{12z}{12} = \dfrac{108}{12}$
    $z = 9$ (**Solving linear equations**)

14. $\dfrac{x}{5} = 9$
    $5\left(\dfrac{x}{5}\right) + (9)5$
    $x = 45$ (**Solving linear equations**)

15. $3x + 18 = 33$
    $3x + 18 - 18 = 33 - 18$
    $3x = 15$
    $x = 5$ (**Solving linear equations**)

16. $5y - 12 = 28$
$5y - 12 + 12 = 28 + 12$
$5y = 40$
$y = 8$ **(Solving linear equations)**

17. $13 = \dfrac{z}{3} - 1$

$3(13) = 3\left(\dfrac{z}{3} - 1\right)$

$39 = z - 3$

$z = 42$ **(Solving linear equations)**

18. $-1 = 11 - 2x$
$-1 - 11 = 11 - 2x - 11$
$-12 = -2x$
$6 = x$ **(Solving linear equations)**

19. $-8 = \dfrac{y}{4} + 7$

$-8 - 7 = \dfrac{y}{4} + 7 - 7$

$-15 = \dfrac{y}{4}$

$y = -60$ **(Solving linear equations)**

20. $6 = 32 + 4x$
$6 - 32 = 32 + 4x - 32$
$-26 = 4x$
$x = \dfrac{-26}{4} = \dfrac{-13}{2} = -6\dfrac{1}{2}$ **(Solving linear equations)**

21. $3(4x - 5) = -3$ Distribute.
$12x - 15 = -3$
$12x = 12$
$x = 1$ **(Solving linear equations)**

22. $10 - 4(3x - 1) = 25$ Distribute.
$10 - 12x + 4 = 25$
$14 - 12x = 25$
$-12x = 11$
$x = \dfrac{-11}{12}$ **(Solving linear equations)**

23. $13 = 3 + 2(x - 7)$
$13 = 3 + 2x - 14$
$13 = -11 + 2x$
$24 = 2x$
$x = 12$ **(Solving linear equations)**

24. $8y - 2y = 30$ Combine like terms.
    $6y = 30$
    $y = 5$ **(Solving linear equations)**

25. $8x = -2x + 30$ Move all terms containing x to the same side.
    $8x + 2x = -2x + 30 + 2x$
    $10x = 30$
    $x = 3$ **(Solving linear equations)**

26. $2(x - 5) + 3x = 15$ Distribute.
    $2x - 10 + 3x = 15$ Combine like terms.
    $5x - 10 = 15$
    $5x = 25$
    $x = 5$ **(Solving linear equations)**

27. $17 - 2x = 9 - 3x$
    $17 - 2x + 3x = 9 - 3x + 3x$
    $17 + x = 9$
    $17 + x - 17 = 9 - 17$
    $x = -8$ **(Solving linear equations)**

28. $\dfrac{x}{2} + \dfrac{x}{3} = 25$ Multiply both sides by the LCD, 6.

    $6\left(\dfrac{x}{2} + \dfrac{x}{3}\right) = 25(6)$ Distribute.
    $3x + 2x = 150$
    $5x = 150$
    $x = 30$ **(Solving linear equations)**

29. $\dfrac{y}{3} = \dfrac{y}{2} - 2$ Multiply both sides by the LCD, 6.

    $6\left(\dfrac{y}{3}\right) = 6\left(\dfrac{y}{2} - 2\right)$ Distribute.
    $2y = 3y - 12$
    $-y = -12$
    $y = 12$ **(Solving linear equations)**

30. $\dfrac{6}{x} + \dfrac{6}{4x} = \dfrac{5}{8}$ Multiply both sides by the LCD, $8x$.

    $8x\left(\dfrac{6}{x} + \dfrac{6}{4x}\right) + 8x\left(\dfrac{5}{8}\right)$ Distribute.
    $8(6) + 2(6) = 5x$
    $48 + 12 = 5x$
    $60 = 5x$
    $x = 12$ **(Solving linear equations)**

31. $\dfrac{2z}{5} = \dfrac{4}{3}$ This can be easily solved by cross multiplication.

$$\cancel{\dfrac{2z}{5}} \bcancel{\dfrac{4}{3}}$$

$3(2z) = 5(4)$

$6z = 20$

$z = \dfrac{20}{6} = \dfrac{10}{3}$ **(Solving linear equations)**

32. $\dfrac{5x}{6} = \dfrac{4x-3}{3}$ Cross multiply.

$3(5x) = 6(4x - 3)$

$15x = 24x - 18$

$-9x = -18$

$x = 2$ **(Solving linear equations)**

33. $\sqrt{6x + 2} + 6 = 10$ Begin by isolating the radical.

$\sqrt{6x + 2} = 4$ Square both sides.

$\left(\sqrt{6x + 2}\right)^2 = 4^2$

$6x + 2 = 16$

$6x = 14$

$x = \dfrac{14}{6} = \dfrac{7}{3}$

Whenever you solve an equation by squaring both sides, it is essential to check the answer, since occasionally, extraneous solutions result from the solution process.

Check:

$\sqrt{6x + 2} + 6 = 10$

$\sqrt{6\left(\dfrac{7}{3}\right) + 2} + 6 = 10$

$\sqrt{16} + 6 = 10$

$4 + 6 = 10$

$10 = 10$ **(Solving linear equations)**

34. $\sqrt{2x} + 8 = 3$ Isolate the radical.

$\sqrt{2x} = -5$ Square both sides.

$\left(\sqrt{2x}\right)^2 = (-5)^2$

$2x = 25$

$x = \dfrac{25}{2}$ Now, the solution must be checked.

Check:

$\sqrt{2x} + 8 = 3$

$\sqrt{2\left(\dfrac{25}{2}\right)} + 8 = 3$

$\sqrt{25} + 8 = 3$

$$5 + 8 = 3$$

$13 \neq 3$. The solution does not check, so there are no solutions to the given equation. (**Solving linear equations**)

35. $\sqrt{-3a} = \sqrt{-5a - 2}$ Square both sides.
$$\left(\sqrt{-3a}\right)^2 = \left(\sqrt{-5a - 2}\right)^2$$
$$-3a = -5a - 2$$
$$2a = -2$$
$a = -1$ This solution can easily be checked. (**Solving linear equations**)

36. $2z + 8 = 18$ (**Setting up word problems**)

37. $\left(\dfrac{z}{3}\right) = 3 + y$ (**Setting up word problems**)

38. $5 + 2w = 14$ (**Setting up word problems**)

39. $6(x - 4) = 3(x + 2)$ (**Setting up word problems**)

40. $7x = 1 + \left(\dfrac{x}{3}\right)$ (**Setting up word problems**)

41. Let $N =$ the number. Then, we have
$$12N - 4 = 6N + 8$$
$$6N = 12$$
$N = \dfrac{12}{6} = 2$. The number is 2. (**Solving word problems**)

42. Let $N =$ the number. Then,
$$3 + 2N = 49$$
$$2N = 46$$
$N = 23$. The number is 23. (**Solving word problems**)

43. Let $N =$ the smallest integer. Then,
$N + 1 =$ the next integer and
$N + 2 =$ the largest integer. We have, then,
$$N + (N + 1) + (N + 2) = 105$$
$$3N + 3 = 105$$
$$3N = 102$$
$N = \dfrac{102}{3} = 34$. The smallest integer is 34. (**Solving word problems**)

44. Let $s =$ the length of the shorter piece.
Then, $36 - s =$ the length of the longer piece. We then obtain
$$2(36 - s) = 2 + 3s$$
$$72 - 2s = 2 + 3s$$
$$70 = 5s$$
$s = 14$. The length of the shorter piece is 14 inches. (**Solving word problems**)

45. Let B = Brian's age. Then,
$3B$ = Janet's age. Furthermore, four years ago, Brian's age would have been $B - 4$, and Janet would have been $3B - 4$. Since four years ago Janet was four times as old as Brian, we have
$3B - 4 = 4(B-4)$
$3B - 4 = 4B - 16$
$B = 12 \quad J = 3 \quad B = 36$
Right now, Janet is 36 years old. (**Solving word problems**)

46. $x = y - z$ for y Add z to both sides.
$y = x + z$ (**Manipulating formulas**)

47. $d = 5e + 2f$ for e Subtract $2f$ from both sides.
$d - 2f = 5e$ Divide by 5.
$e = \dfrac{d - 2f}{5}$ (**Manipulating formulas**)

48. $ax - b = 3c$ for x Add b to both sides.
$ax = 3c + b$ Divide by a.
$x = \dfrac{3c + b}{a}$ (**Manipulating formulas**)

49. $P = 2(L + W)$ for L Distribute.
$P = 2L + 2W$
$P - 2W = 2L$
$L = \dfrac{P - 2W}{2}$ (**Manipulating formulas**)

50. $\dfrac{q}{x} + \dfrac{p}{4x} = 2$ for x Multiply by $4x$.
$4x\left(\dfrac{q}{x} + \dfrac{p}{4x}\right) = 4x(2)$
$4q + p = 8x$
$x = \left(\dfrac{4q + p}{8}\right)$ (**Manipulating formulas**)

51. $3x - 5 > 16$
$3x > 21$
$x > 7$ (**Solving and graphing inequalities**)

52. $8x + 2 < 18$
$8x < 16$
$x < 2$ (**Solving and graphing inequalities**)

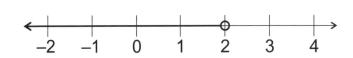

53. $\dfrac{x}{6} - \dfrac{x}{2} \leq 2$ Multiply by 6.
$6\left(\dfrac{x}{6} - \dfrac{x}{2}\right) \leq 2(6)$
$x - 3x \leq 12$
$-2x \leq 12$ Divide by -2 and change the direction of the inequality.
$x \geq -6$ (**Solving and graphing inequalities**)

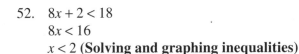

54. $6y + 5 \geq 3(y - 2)$ Distribute.

$6y + 5 \geq 3y - 6$

$3y + 5 \geq -11$

$3y \geq -16$

$y \geq -\dfrac{16}{3}$ Note that at no time did we multiply or divide by a negative number, so the direction of the inequality did not change. (**Solving and graphing inequalities**)

55. One way to easily find five solutions to the given equation is to select values for x and compute the corresponding values for y. For example, if $x = 1$, we have $2(1) + y = 6$ or $y = 4$. Thus, one solution is $x = 1$, $y = 4$. Similarly, we can find other solutions such as $x = 0$, $y = 6$; $x = 2$, $y = 2$; $x = 3$, $y = 0$, and $x = 4$, $y = -2$. (**Solving linear equations in unknowns**)

56. If $x = 5$, we can solve for y to find the corresponding value, as follows:

$5x - 2y = 7$

$5(5) - 2y = 7$

$25 - 2y = 7$

$-2y = -18$

$y = 9$ (**Solving linear equations in unknowns**)

57. If we let $x = 0$, we can compute $y = -2$. Similarly, if $y = 0$, $x = 3$. Another solution is $x = 6$, $y = 2$. (**Solving linear equations in unknowns**)

58. If $y = 4$, we have $2x = 7(4) - 14$ or $2x = 14$. Thus, $x = 7$. (**Solving linear equations in unknowns**)

Grade Yourself

Circle the numbers of the questions you missed, then fill in the total incorrect for each topic. If you answered more than three questions incorrectly, you need to focus on that topic. (If a topic has fewer than three questions and you had at least one wrong, we suggest you study that topic also. Read your textbook or a review book, or ask your teacher for help.)

Subject: Linear Equations and Inequalities

Topic	Question Numbers	Number Incorrect
Classifying equations	1, 2, 3, 4, 5, 6, 7, 8, 9, 10	
Solving linear equations	11, 12, 13, 14, 15, 16, 17, 18, 19, 20, 21, 22, 23, 24, 25, 26, 27, 28, 29, 30, 31, 32, 33, 34, 35	
Setting up word problems	36, 37, 38, 39, 40	
Solving word problems	41, 42, 43, 44, 45	
Manipulating formulas	46, 47, 48, 49, 50	
Solving and graphing inequalities	51, 52, 53, 54	
Solving linear equations in unknowns	55, 56, 57, 58	

Functions and Graphs

3

Brief Yourself

The Cartesian coordinate system, which is used in plotting graphs of functions as well as in proving theorems in analytic geometry, consists of two perpendicular lines called the *x-axis* and the *y-axis* on a plane. The *x*-axis is normally horizontal, and the *y*-axis is vertical. The point where the axes intersect is given the coordinates $(0, 0)$, and is called the *origin*. A point P on the *x*-axis that is a units to the right of the origin is given the coordinates $(a, 0)$. If a point P is a units to the left of the origin it is given coordinates $(-a, 0)$. A point P on the *y*-axis that is a units above the origin is given the coordinates $(0, a)$; if the point is a units below the origin, it is given coordinates $(0, -a)$. If a point P is on neither coordinate axis, drop perpendiculars from the point P to the *x*-axis and to the *y*-axis. Let $(a, 0)$ be the coordinates of the foot of the perpendicular to the *x*-axis, and let $(0, b)$ be the coordinates of the foot of the perpendicular to the *y*-axis. Then P is assigned the coordinates (a, b). If we are given the coordinates (a, b), it is easy to find the unique point that corresponds to those coordinates.

A Cartesian coordinate system establishes a one-to-one correspondence between the set of pairs of real numbers (x, y) and the set of points in a plane. Figures in the plane can be associated with equations relating x to y. For example, it you have a linear equation such as $y = 2x + 3$, the points $(-1,1)$ and $(1, 5)$ are found by first plugging $x = -1$ and then $x = +1$ into the formula for y. Having plotted these two points, the rest is easy. Simply take a ruler and pass a straight line through the two points. Every point on this line will correspond to a number pair (x, y) that satisfies the equation, and every number pair that satisfies the equation will correspond to a point on the line.

If (x_1, y_1) and (x_2, y_2) are two points on the equation of a line, we define $m = \dfrac{y_2 - y_1}{x_2 - x_1}$ to be the *slope of the line*. Intuitively, the slope represents the *steepness* of the line. A horizontal line will have a slope of 0. A line that increases from left to right will have a positive slope; a line that decreases from left to right will have a negative slope.

There are various forms in which you can put the equation of a line. If you are given that the *y*-intercept is b and the slope is m, then

$$y = mx + b$$

(which is called the *slope-intercept form*) is convenient. If you are told that the slope of the line is m and the line passes through the point (x_1, y_1), then

$$y - y_1 = m(x - x_1)$$

(which is called the *point-slope form*) is convenient. If you are simply given the equation of the line, then sometimes the form

$$Ay + Bx = C$$

(which is called the *standard form*) is convenient.

The distance between two points (x_1, y_1) and (x_2, y_2) in a Cartesian coordinate system can be easily calculated. By the Pythagorean theorem, it is seen to be

$$d = \sqrt{(x_2-x_1)^2 + (y_2-y_1)^2}$$

The equation of a circle with center at the origin is

$$x^2 + y^2 = r^2$$

where r is the radius of the circle. The equation of a circle of radius r and center (h, k) is

$$(x - h)^2 + (y - k)^2 = r^2$$

Other curves that can be represented by quadratic equations in x and y are the parabola, hyperbola, and ellipse. If you have a point (called the *focus*) and a line (called the *directrix*), then consider the focus of all points such that the distance from the focus is e times the distance from the directrix. The ratio e is called the *eccentricity*. If $0 < e < 1$, the figure that results is an ellipse. If $e = 1$, the figure is a parabola. If $e > 1$, then the figure is a hyperbola. By rotation or translation of the axes, these equations can be put in one of the following three forms:

Parabola with focus at $(0, p)$ and directrix $y = -p$: $4py = x^2$

Ellipse with focus at $(\sqrt{(a^2+b^2)}, 0)$ and directrix at $x = \dfrac{a}{e}$: $\left(\dfrac{x}{a}\right)^2 + \left(\dfrac{y}{b}\right)^2 = 1$

$$\sqrt{a^2 - b^2}$$

Hyperbola with focus at $(\sqrt{(a^2+b^2)}, 0)$ and directrix at $x = \dfrac{a}{e}$: $\left(\dfrac{x}{a}\right)^2 - \left(\dfrac{y}{b}\right)^2 = 1$

If you are given a general quadratic function of x and y in the form

$$Ax^2 + Bxy + Cy^2 + Dx + Ey + F = 0$$

then it is a parabola if $B^2 - 4AC = 0$, an ellipse if $B^2 - 4AC < 0$, and a hyperbola if $B^2 - 4AC > 0$. If $B = 0$ and $A = C$, it is a circle.

A *function* is a rule that assigns to each real number in a set called the *domain* of the function exactly one real number in another set called the *range* of the function.

Examples of functions are $y = x^2$, which has the set of all real numbers as its domain and the set of nonnegative real numbers as its range; $y = \sqrt{x}$, which has the set of nonnegative real numbers as its domain and the set of all nonnegative real numbers as its range; $y = \sin x$, which has the set of all real numbers as its domain and the real numbers between -1 and $+1$ (inclusive) as its range; $y = 1$, which has the set of all real numbers as its domain and the single real number 1 as its range; and $y = \dfrac{1}{(x^2-1)}$, which has all the real numbers except $x = +1$ and $x = -1$ as its domain and all the reals less than or equal to -1 together with all the reals greater than 0 as its range.

A function $y = f(x)$ can be represented as a graph in the (x, y) plane, where the graph passes through all those points and only those points that have the form $(x, f[x])$. A curve in the (x, y) plane will represent a function if all vertical lines (that is, lines of the form $x = c$) intersect the graph in at most one point. This is called the *vertical line test*. For example, the curve $x^2 + y^2 = 1$ fails the vertical line test, because the line $x = 0$ intersects it at both $y = +1$ and $y = -1$. Thus, $x^2 + y^2 = 1$ does not represent a function.

For functions other than linear functions, it may be quite difficult to plot the function. It is always easy to find a few points on the graph by simply calculating the value the function takes for various convenient values of x, but to determine the graph between these points may be tricky. In calculus tools are developed that help in this.

If $y = f(x)$ is a given function, the y intercept of the graph of $f(x)$ is the value of y such that $y = f(0)$. The x-intercepts of the graph of $f(x)$ consist of those values of x (and there may be many of them!) such that $f(x) = 0$. Thus, to find the y-intercept of a function, set $x = 0$ and to find the x-intercepts, set $f(x) = 0$.

Certain properties of a function may be reflected in symmetry properties of its graph. For example, if $f(x) = f(-x)$ for all x, it can be seen that the graph will be symmetric about the y-axis. Functions with this property are called *even functions*. More generally, if $f(a + x) = f(a - x)$ for all x, then the graph will be symmetric about the vertical line $x = a$.

If $f(x) = y$ implies that $f(y) = x$, then if (a, b) is on the graph, (b, a) is also on the graph. This means the graph is symmetric about the line $y = x$. An example of a function with this property is $y = \dfrac{1}{x}$.

If $y = f(x)$ implies that $-y = f(-x)$, then the graph of the function is symmetric in the origin. That is, if you rotate the graph through 180 degrees about the origin, it is carried onto itself. Functions with this property are called *odd functions*. Examples of odd functions are $y = \sin x$ and $y = x^3$. Note that odd functions must have graphs that go through the origin.

Functions can be added, subtracted, multiplied, and divided as follows:

$(f + g)(x) = f(x) + g(x)$

$(f - g)(x) = f(x) - g(x)$

$(fg)(x) = f(x)g(x)$

$\left(\dfrac{f}{g}\right)(x) = \dfrac{f(x)}{g(x)}$ (Note: the quotient is undefined if $g(x) = 0$.)

There is another operation between functions called *functional composition*, which is written $f \circ g$. It is defined by the following equation:

$$(fog)(x) = f(g(x))$$

Functional composition is only defined if x is in the domain of g and $g(x)$ is in the domain of f. f is called the *inverse of function g* if

$$(fog)(x) = (gof)(x) = x$$

The function g has an inverse if its graph passes the *horizontal line test*, which requires that each line of the form $y = a$ cuts the graph $(x, g[x])$ at most once. For example, the function $g(x) = x^2$ fails the horizontal line test because the line $y = 1$ cuts the graph at the points $(-1, 1)$ and $(1, 1)$. To define an inverse for $g(x)$, we must restrict its domain to the nonnegative real numbers. Similarly $g(x) = \sin x$ fails the horizontal line test, but we can restrict its domain to values of x between $\frac{-\pi}{2}$ and $\frac{+\pi}{2}$ and then define an inverse function.

Examples of inverse functions are $\arcsin(y)$, $\arccos(y)$, $\arctan(y)$, which are inverse functions to sine, cosine, and tangent respectively. Another example is $\log(y)$ which is the inverse function to $y = 10^x$. The function $y = \frac{1}{x}$ is its own inverse. You can find the inverse of a function by simply interchanging x and y in the equation that defines the function and then solving for y. For example, the inverse of the linear function

$$y = g(x) = 2x + 1$$

is found by solving $x = 2y + 1$ and getting

$$y = f(x) = \frac{(x-1)}{2}$$

You can check that *fog* carries each value of x onto itself.

Test Yourself

1. Draw a Cartesian coordinate system and plot the following points: A (2, 3), B (−5, 7), C (−3, −3), and D (−4, 2).

2. Calculate the distance of each of the points given in problem 1 from the origin.

3. Find the equation of the line passing through the points (−4, 1) and (0, 5). Find the equation of the line passing through (−2, 2) and (0, 0).

4. Two lines are perpendicular to one another if the product of their slopes is −1. Find the equation of a line through the origin that is perpendicular to the line passing through (0, 2) and (5, 0).

5. Find the area of the triangle whose vertices are (−3, 1), (0, 4), and (4, −1).

6. Find the equation of the ellipse whose center is the origin whose axes are parallel to the x- and y-axes and which passes through the point (1, 2) and the point (2, 0). What is the eccentricity of this ellipse?

7. Find the equation of the circle whose center is at (1, 2) and whose radius is 5 units.

8. Find the equation of the parabola with focus at (0, 4) and directrix $y = -4$.

9. Graph the hyperbola $\left(\frac{x}{2}\right)^2 - \left(\frac{y}{3}\right)^2 = 1$.

10. Consider the conic section defined by $x^2 + 2xy + Cy^2 + 7x - 3y + 29 = 0$. For what values of C is this an ellipse? A parabola? A hyperbola?

11. Find the points of intersection of the line $y = 2x + 3$ and the circle $x^2 + y^2 = 9$.

12. What are the x- and y-intercepts of the curve $x^2 + 2xy + 3y^2 + 2x - 4y + 1 = 0$?

13. Find the points of intersection of the circle $x^2 + y^2 = 1$ and the parabola $y = 2x^2 - 1$.

14. Find the equation of the line that is tangent to the circle $(x - 1)^2 + y^2 = 4$ at the point $(2, \sqrt{3}\,)$.

15. For what values of h and k does the circle $(x - h)^2 + (y - k)^2 = 4$ pass through the point $(1, 2)$?

16. Sketch the curve $y = \dfrac{(x + 1)}{(x - 1)}$. For what values of x is the function undefined? What happens to y as x becomes large and positive? What happens to y as x becomes large and negative?

17. Sketch the curve $y = x^3 - 2x$. What are the x-intercepts of the curve? What are its y-intercepts?

18. Suppose that G is the set of all points (x, y) that satisfy $P(x, y) = 0$, and H is the set of all points that satisfy $Q(x, y) = 0$. What set of points will satisfy $P(x, y)Q(x, y) = 0$?

19. Sketch the curve $y = \dfrac{x^2 + 1}{x - 2}$. What happens to y as x becomes large and positive? What happens to y as x becomes large and negative?

20. Suppose -1 and $+1$ are the x-intercepts of $y = f(x)$, and -2 is the sole x-intercept of $y = g(x)$. What can you say about the x-intercepts of $y = f(x)g(x)$? What can you say about the y-intercept of $y = f(x)g(x)$?

21. Suppose M is the maximum value of $y = f(x)$, and its x-intercepts are -1 and $+1$. What are the x-intercepts and the maximum value of $y = 3f(x + 2)$?

22. Sketch the curve $y = \dfrac{x^2 - 3x - 4}{x + 1}$. What are the x- and y-intercepts? Where is the function undefined?

23. Sketch the curve $y = \dfrac{x^2 + 5x + 4}{x + 1}$. Where are the intercepts? Where is the function undefined?

24. Suppose $f(x)$ is a polynomial of degree 3. What is the maximum possible number of x-intercepts? What is the minimum possible number of x-intercepts?

25. A function is called a *rational function* if it is the ratio of two polynomials. Is there any rational function that does not have a y-intercept?

26. Recall that if a function $y = f(x)$ has the property that $f(-x) = -f(x)$ for all x, it is called an *odd function*. If it has the property that $f(x) = f(-x)$ for all x, it is called an *even* function. Is there any function that is both odd and even?

27. Classify the following functions as odd, even, or neither:

    (a) $y = x^2 + 1$

    (b) $y = x^3 + x$

    (c) $y = x + 1$

    (d) $y = (x^2 + 1)^3$

28. If $f(x)$ is an even function and $g(x)$ and $h(x)$ are both odd functions, classify the following functions as odd, even, or neither:

    (a) $y = f(x)g(x)$

    (b) $y = g(x)h(x)$

    (c) $y = g(x) + h(x)$

    (d) $y = f(x) + g(x)$

29. If $f(x)$ is an arbitrary function defined for all values of x, classify the following functions as always odd, always even, or neither:

    (a) $y = f(x) + f(-x)$

    (b) $y = f(x) - f(-x)$

    (c) $y = f(x)f(-x)$

    (d) $y = \left[\dfrac{f(x) + f(-x)}{f(x) - f(-x)}\right]$

30. Prove that every function $f(x)$ can be written as the sum of an odd function and an even function.

31. If $f(x) = x^2 + 1$ and $g(x) = 2x - 1$, write out *fog* and *gof* (remember that *fog* is *f* composed with *g*, or $f(g[x])$).

32. If f is an odd function, and g is an even function, what can you say about *fog*? What can you say about *gof*? If h is an odd function, what can you say about *foh*?

33. If $f(x) = \dfrac{ax + b}{cx + d}$, what is the inverse function to $f(x)$? Verify that if $g(x)$ is the inverse you have found, then $g(f[x])$.

34. If $f(x) = \dfrac{ax + b}{cx + d}$, what conditions must the constants a, b, c, d satisfy if f is even? What conditions must they satisfy if f is odd?

35. The graph of $f(x)$ is symmetric about the line $x = d$ if $f(d + x) = f(d - x)$. If $f(x)$ is symmetric about $x = d$, what does this mean about the function $g(x) = f(x + d)$?

36. Suppose $f(x) = x^2 + 5x + 1$. Calculate $f(x + h) - f(x)$ and $\dfrac{[f(x + h) - f(x)]}{h}$. What happens to the latter function if $h = 0$?

37. Suppose that c is one of the x-intercepts of the function $x = f(x)$. Let $g(x) = 2x + 3$. Can you calculate one of the x-intercepts of $(fog)(x)$ in terms of c?

38. Suppose $f(x) = \dfrac{[x-1]}{[x+1]}$. Find a function $g(x)$ such that $(gof)(x) = x^2$.

39. Which of the following statements is always true?

    (a) $(f + g)oh = foh + goh$

    (b) $ho(f + g) = hof + hog$

    (c) $(fg)oh = foh(goh)$

    (d) $fo(gh) = (fog)(foh)$

40. Suppose $f(x) = mx + b$, and $g(x) = nx + c$. What conditions must the constants m, n, b, c satisfy if *fog* = *gof*?

41. The fuel efficiency E of a 3000-pound car (in miles per gallon) is a function of its average speed v. The function is approximately $E = 34 - \dfrac{(v-40)^2}{15}$. At what speed is the greatest fuel economy achieved? At what two speeds will the fuel economy be 19 miles per gallon?

42. A company makes boxes of various sizes. All the boxes are twice as long as they are wide and have a height 2 inches less than their width. Write a function that gives the volume of the box as a function of its width.

43. Two caravans leave the same oasis at the same time. One is heading due north at a constant speed of 1.5 miles per hour. The other is heading due west at a constant speed of 2 miles per hour. Write a function describing their distance apart as a function of the number of hours since they left the oasis.

44. A box with a square base and no top contains 8 cubic meters. If the material for the base costs $10 per square meter and the material for the sides costs $5 per square meter, write the formula for the cost of the materials for the box as a function of the width of the box.

45. A company manufactures and sells bicycles. If x is the number of bicycles it builds each week, its cost each week is $1000 + 50x$. If y is the price it charges for each bicycle, then the number it sells each week is $100 - \dfrac{y}{10}$. Express the company's gross weekly profit as a function of the price it charges for its bicycles.

46. The kinetic energy (E) and momentum (M) of a particle of mass m moving with velocity v is given in Newtonian mechanics to be

    $E = \dfrac{mv^2}{2}$ and $M = mv$. Express the mass and velocity as a function of the momentum and kinetic energy alone.

47. The baseball diamond is a square 90 feet on a side. If a base runner leaves first base at time $t = 0$ with a velocity of 30 feet per second, express his distance from home plate as a function of t.

48. Rapacious County charges out-of-state motorists for speeding on a sliding scale of $50 plus $20 for every mile over the speed limit. The speed limit is 35 miles per hour within 3 miles of any school or other occupied building. Give a function that expresses fine (F) as a function of speed (S), and give the function that expresses speed (S) as a function of fine (F).

49. In the state of Euphoria, the state income tax takes nothing from the first $10,000 you make, 5% of each dollar you make over $10,000 and less than $30,000, 10% of each dollar you make over $30,000 and less than $100,000, and 20% of each dollar you make over $100,000. Express the tax due as a function of the amount you make each year. A certain candidate for governor proposes replacing this complex schedule with a flat tax of 6% on every dollar you make. Who would do better under this scheme and who would do worse?

Check Yourself

1.

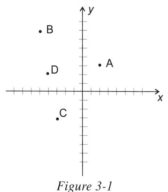

Figure 3-1

(Graphing functions)

2. For $A(2, 3)$, we compute $d = \sqrt{(2-0)^2 + (3-0)^2} = \sqrt{4} + 9 = \sqrt{13}$. Thus, $|OA| = \sqrt{13}$. Similarly, $|OB| = \sqrt{74}$. $|OC| = 3\sqrt{2}$. $|OD| = 2\sqrt{5}$. **(Using functions in geometry)**

3. The first line has equation $y = x + 5$. The second line has equation $y = -x$. **(Using functions in geometry)**

4. The slope of the line through $(0, 2)$ and $(5, 0)$ is $\dfrac{-2}{5}$, so we are looking for a line with slope $\dfrac{5}{2}$. The line through the origin with this slope is $y = \dfrac{5x}{2}$, or $2y = 5x$. **(Using functions in geometry)**

5. The slope of the line through (–3, 1) and (0, 4) is 1, and its equation is $y = x + 4$. The perpendicular to this side from the point (4, –1) will be $y = -x + 3$. These lines intersect at $\left(\frac{-1}{2}, \frac{7}{2}\right)$. Now apply the distance formula and the formula $A = \frac{bh}{2}$ to find that the area of the triangle is $\frac{27}{2}$. **(Using functions in geometry)**

6. The equation must be of the form $\frac{x^2}{a^2} + \frac{y^2}{b^2} = 1$. Direct substitution of the known points on the curve shows that $a^2 = 4$ and $b^2 = \frac{16}{3}$. The major semi-axis is along the *y*-axis. The eccentricity is $\frac{c}{a}$, which is $\frac{1}{2}$ in this case. **(Using functions in geometry)**

7. The equation is $(x - 1)^2 + (y - 2)^2 = 25$. **(Using functions in geometry)**

8. $x^2 + (y - 4)^2 = (y + 4)^2$, which simplifies to $x^2 = 16y$. **(Using functions in geometry)**

9.

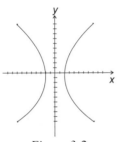

Figure 3-2

(Graphing functions)

10. If $C > 1$, it is an ellipse, if $C = 1$, it is a parabola, and if $C < 1$, it is a hyperbola. **(Graphing functions)**

11. The points of intersection are (0, 3) and $\left(\frac{-12}{5}, \frac{-9}{5}\right)$. This can be found by substituting $y = 2x + 3$ into $x^2 + y^2 = 0$:

    $x^2 + (2x + 3)^2 - 9$
    $x^2 + (4x^2 + 12x + 9) = 9$
    $5x^2 + 12x + 9 = 9$
    $5x^2 + 12x = 0$
    $x(5x + 12) = 0$
    $x = 0$. Then, use $y = 2x + 3$ to find the corresponding values of *y*. **(Using functions in geometry)**

12. By first setting $x = 0$, and then $y = 0$, we find that the *x*-intercepts are $x = -1$ and the *y*-intercepts are $y = 1$ and $y = \frac{1}{3}$. **(Using functions in geometry)**

13. $(0, -1), \left(\frac{\sqrt{3}}{2}, \frac{1}{2}\right)$, and $-\left(\frac{\sqrt{3}}{2}, \frac{1}{2}\right)$. This solution can be found by substituting $y = 2x^2 - 1$ into $x^2 + y^2 = 1$ and solving the resulting equation. **(Using functions in geometry)**

14. The center of the circle is at $(1, 0)$, so the slope of the radius to the given point is $\frac{\sqrt{3}}{2}$, and we want the slope of the tangent line to be $\frac{-2}{\sqrt{3}}$. The equation of the tangent line is $y - \sqrt{3} = \frac{-2}{\sqrt{3}} (x - 2)$. (**Using functions in geometry**)

15. All values of h and k that satisfy $(h - 1)^2 + (k - 2)^2 = 4$. (**Using functions in geometry**)

16. See Figure 3-3. The function is undefined at $x = 1$. When x gets large and positive, the curve approaches the line $y = 1$ from above. When x is large and negative, the curve approaches $y = 1$ from below.

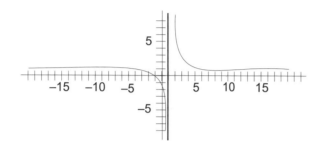

Figure 3-3

(**Using functions in geometry**)

17. See Figure 3-4. The x-intercepts are $x = 0$, $x = \sqrt{2}$ and $x = -\sqrt{2}$. The y-intercept is $y = 0$.

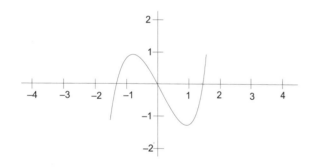

Figure 3-4

(**Using functions in geometry**)

18. The union of the two sets G and H. (**Using functions in geometry**)

19. See Figure 3-5. As x becomes large and positive or large and negative the curve approaches the line $y = x + 2$. (**Using functions in geometry**)

20. The x-intercepts will be -2, -1, and $+1$. The y-intercept will be the product of the y-intercepts of $f(x)$ and $g(x)$. (**Using functions in geometry**)

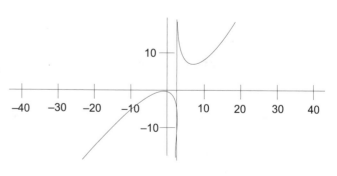

Figure 3-5

21. The maximum value will be 3M, and the x-intercepts will be −1 and −3. (**Graphing functions**)

22. See Figure 3-6. This function is the line $y = x - 4$, except that it is undefined when $x = -1$. The x-intercept is $x = 4$. The y-intercept is $y = -4$.

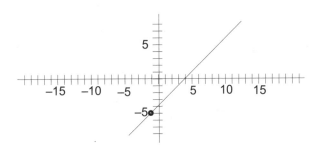

Figure 3-6

(**Graphing functions**)

23. See Figure 3-7. This function is undefined when $x = -1$. The x-intercept is at $x = -4$. The y-intercept is $y = 4$.

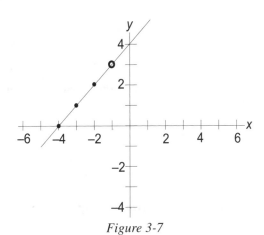

Figure 3-7

(**Graphing functions**)

24. The maximum possible number is 3, and the minimum possible number is 1. (**Graphing functions**)

25. $y = \dfrac{1}{x}$ is a rational function and does not have a y-intercept. (**Functional composition and other properties of functions**)

26. If $f(x) = f(-x) = -f(x)$, then $2f(x) = 0$, and so only the function that is identically 0 is both odd and even. (**Functional composition and other properties of functions**)

27. (a) even (b) odd (c) neither (d) even (**Graphing functions**)

28. (a) odd (b) even (c) odd (d) neither (assuming neither f nor g is the zero function) (**Graphing functions**)

29. (a) always even (b) always odd (c) always even (d) always odd (**Graphing functions**)

30. $f(x) = \dfrac{(f(x) + f(-x))}{2} + \dfrac{(f(x) - f(-x))}{2}$. (**Functional composition and other properties of functions**)

31. $f \circ g(x) = (2x - 1)^2 + 1 = 4x^2 - 4x + 2.$ $g \circ f(x) = 2(x^2 + 1) - 1 = 2x^2 + 1.$ (**Functional composition and other properties of functions**)

32. $f \circ g$ is an even function. $g \circ f$ is also an even function. $f \circ h$ is an odd function. (**Functional composition and other properties of functions**)

33. The inverse function is $\dfrac{[dx - b]}{[-cx + a]}$. (**Functional composition and other properties of functions**)

34. The function is even if and only if $ad = bc$. This implies, by a little algebra, that the function is the constant function $y = \dfrac{a}{c}$. The function will be odd if and only if $bd = 0$ and $ac = 0$. d and c cannot both be zero or the function is undefined. Therefore we must either have $f(x) = \dfrac{ax}{d}$ or $f(x) = \dfrac{b}{cx}$. (**Functional composition and other properties of functions**)

35. $g(x) = g(-x)$ so g is an even function. (**Functional composition and other properties of functions**)

36. $f(x + h) - f(x) = 2hx + h^2 + 5h.$ $\dfrac{[f(x + h) - f(x)]}{h} = \dfrac{[2hx + h^2 + 5h]}{h}$. If $h = 0$ this function is undefined. If h is very close to 0, this function is very close to $2x + 5$. (**Functional composition and other properties of functions**)

37. $x = \dfrac{(c - 3)}{2}$ will be one of the x-intercepts of $f \circ g$. (**Functional composition and other properties of functions**)

38. $g(x) = \left[\dfrac{(1 + x)}{(1 - x)}\right]^2$ (**Functional composition and other properties of functions**)

39. (a) always true (b) not always true (c) always true (d) not always true (**Functional composition and other properties of functions**)

40. $c(m - 1) = b(n - 1)$. (**Functional composition and other properties of functions**)

41. The greatest fuel economy is achieved at $v = 40$ mph. The efficiency is 19 miles per gallon at 25 mph and at 55 mph. (**Functional composition and other properties of functions**)

42. $V = (2W)(W)(W - 2) = 2W^3 - 4W^2$ gives the volume in cubic inches. (**Functional composition and other properties of functions**)

43. $d = 2.5t$ gives their distance in miles if t is the time since they left the oasis in hours.
$d^2 = (2t)^2 + (1.5t)^2 = 6.25t^2$
Thus, $d = 2.5t$. (**Functional composition and other properties of functions**)

44. The cost in dollars is $10W^2 + \dfrac{160}{W}$ if W is the width in meters. (**Functional composition and other properties of functions**)

45. Assuming the company plans production to equal the number of bicycles sold each week, the gross profit will be $105y - \dfrac{y^2}{10} - 6000$.

Profit = Revenue – Cost

Revenue = (Number Sold)(Price) = $(100 - \dfrac{y}{10})(y) = 100y - \dfrac{y^2}{10}$

Cost = $1000 + 50x$

If the company builds as many bikes as it sells, $x = 100 - \dfrac{y}{10}$, so

Cost = $1000 + 50(100 - \dfrac{y}{10}) = 1000 + 5000 - 5y = 6000 - 5y$.

Then, Profit = $100y - \dfrac{y^2}{10} - (6000 - 5y)$

$\qquad\qquad = 100y - \dfrac{y^2}{10} - 6000 - 5y$

$\qquad\qquad = 105y - \dfrac{y^2}{10} - 6000$

(Functional composition and other properties of functions)

46. $v = \dfrac{2E}{M}$ and $m = \dfrac{M^2}{2E}$

If $M = mv$, we have $v = \dfrac{M}{m}$. Then, $E = \dfrac{mv^2}{2} = \dfrac{m\left(\dfrac{M}{m}\right)^2}{2} = \dfrac{mM^2}{2m^2} = \dfrac{M^2}{2m}$. From $E = \dfrac{M^2}{2m}$, we obtain $m = \dfrac{M^2}{2E}$.

Similarly, if $M = mv$, we have $m = \dfrac{M}{V}$. Then, $E = \dfrac{mv^2}{2} = \dfrac{\left(\dfrac{m}{v}\right)v^2}{2} = \dfrac{Mv}{2}$. From $E = \dfrac{Mv}{2}$, we obtain $V = \dfrac{2E}{M}$.

(Applications of functions)

47. $d = 30\sqrt{(9 + t^2)}$ gives the distance from home plate in feet, provided t is between 0 and 3 seconds.

$d^2 = 90^2 + (30t)^2$

$d = \sqrt{90^2 + 30^2 t^2} = 30\sqrt{(9 + t^2)}$ When $t = 3$ seconds the runner reaches second base. **(Applications of functions)**

48. $F = 50 + 20(S - 35)$ if $S \geq 35$, and $F = 0$ otherwise, S being the speed in miles per hour, expressed as an integer. $S = 35 + \dfrac{[F - 50]}{20}$, provided F is not 0. If F is 0, all you know about S is that it is 35 or less. **(Applications of functions)**

49. $T(x) = 0$ for $x \leq 10,000$; $T(x) = \dfrac{(x - 10,000)}{20}$ for $x > 10,000$ and $x \leq 30,000$; $T(x) = 1,000 + \dfrac{(x - 30,000)}{10}$ for $x > 30,000$ and $x \leq 100,000$; $T(x) = 8,000 + \dfrac{(x - 100,000)}{5}$ for $x > 100,000$. The candidate's scheme would make the tax .06x for all x. This function is higher than the existing tax for all x up to $x = 50,000$, and for larger x it is less. So anyone making more than \$50,000 a year would benefit from the candidate's proposal, and anyone making less than that would have his or her taxes raised. **(Applications of functions)**

Grade Yourself

Circle the numbers of the questions you missed, then fill in the total incorrect for each topic. If you answered more than three questions incorrectly, you need to focus on that topic. (If a topic has fewer than three questions and you had at least one wrong, we suggest you study that topic also. Read your textbook or a review book, or ask your teacher for help.)

Subject: Functions and Graphs

Topic	Question Numbers	Number Incorrect
Graphing functions	1, 9, 10, 21, 22, 23, 24, 27, 28, 29	
Using functions in geometry	2, 3, 4, 5, 6, 7, 8, 11, 12, 13, 14, 15, 16, 17, 18, 19, 20	
Functional composition and other properties of functions	25, 26, 29, 30, 31, 32, 33, 34, 35, 36, 37, 38, 39, 40, 41, 42, 43, 44, 45	
Applications of functions	46, 47, 48, 49	

Exponents and Logarithms

4

 Brief Yourself

If you multiply a quantity, b, times itself n times, the result is called "b to the nth power" and is written b^n. The number b is called the *base* and the number n is called the *exponent*. The definition of positive integral exponents can be readily extended to exponents that are negative integers, fractions, and even to exponents that are irrational numbers.

There are seven fundamental laws of exponents:

1. $(b^x)(b^y) = b^{x+y}$

2. $\dfrac{(b^x)}{(b^y)} = b^{x-y}$ (b not equal to zero)

3. If b is not -1, 0, or $+1$, then $b^x = b^y$ if and only if $x = y$.

4. $(ab)^x = (a^x)(b^x)$

5. $\left(\dfrac{a}{b}\right)^x = \dfrac{a^x}{b^x}$ (b not equal to 0)

6. If x is not 0 and a and b are both greater than 0, then $a^x = b^x$ if and only if $a = b$.

7. $(b^x)^y = b^{(xy)}$

Based on the laws of exponents, we define $b^{-x} = \dfrac{1}{b^x}$ and that $b^0 = 1$ if b is not equal to 0.

A function of the form $f(x) = ab^x$ is called an *exponential function*. Such functions have many applications in population growth, compound interest, radioactive decay, and other areas.

The inverse functions of exponential functions of the form $f(x) = b^x$ are called *logarithms to the base b*. We write "$\log_b y = x$" if $b^x = y$. All logarithms, whatever their base, obey four fundamental laws of logarithms.

If $b > 0$, $M > 0$, $N > 0$, and b is not equal to 1, then

1. $\log_b (MN) = \log_b M + \log_b N$

2. $\log_b \left(\dfrac{M}{N} \right) = \log_b M - \log_b N$

3. $\log_b M = \log_b N$ if and only if $M = N$

4. $\log_b M^r = r \log_b M$, for any real number r

Common logarithms are logarithms with the base 10 and are usually written as "log" rather than "$\log_{10}$." Another base often used for logarithms is the number e, defined as the limit, as n becomes very large, of $\left[1 + \left(\dfrac{1}{n} \right) \right]^n$. The numerical value of e is 2.718281828... Logarithms to the base e are called *natural logarithms* and are often written "ln" instead of "$\log_e$."

You can use the change of base formula to rewrite logarithms with a particular base as logarithms with another:

$$\log_b M = \frac{\log_a M}{\log_a b}$$

Logarithms are very useful in solving equations in which the unknown is in an exponent. For example:

$$8^x = \sqrt{5}$$

$$8^x = 5^{1/2}$$

$$\log 8^x = \log 5^{1/2}$$

$$x \log 8 = \frac{\log(5)}{2}$$

$$x = \frac{\log 5}{2 \log 8} = \frac{.69897}{1.80618} = .38699$$

Test Yourself

Simplify each of the following expressions.

1. $(-4)^{(-2)}$

2. $5(2)^{(-2)}$

3. $[(2)^{(-1)} + (2)^{(-2)}]^{(-1)}$

4. $(3a^{(-2)})^3$

5. $\dfrac{8n^3 - 4n^{(-2)}}{2(n)^{(-2)}}$

6. $\dfrac{(3^5)\,(9^4)}{(27^4)}$

7. $\dfrac{(b(b)^n)^2}{(b^2)^n}$

8. $\dfrac{(2^{(-1)})}{[2^{(-2)} + 2^{(-3)}]}$

9. $\left(2^{(-1)} + 4^{(-1)}\right)^{(-1)}$

10. $\left(\dfrac{9}{25}\right)^{\frac{1}{2}}$

Evaluate or simplify the expressions in problems 11–19. Throughout, we use "log" for the common logarithm (that is, logarithm to the base 10) and "ln" for the natural logarithm (that is, logarithm to the base e.)

11. $\log(100)$; $\log(10)$; $\log(1)$; $\log\left(\dfrac{1}{100}\right)$.

12. If $\ln(2) = .6931$, what is $\ln(8)$?

13. If $\log(2) = .30103$, what is $\log(5)$? What is $\log(16)$? What is $\log(1.6)$?

14. $\ln(e)$; $\ln(e^3)$; If $\ln(10) = 2.3026$, what is $\ln(100)$? What is $\ln(27.1828)$?

15. Express $\log(M) + 2\log(N) + 2$ as the log of a single quantity.

16. If $\log(M) = a$ and $\log(N) = b$, then what is $\log(M^2 N)$?

17. Express $\log \pi + 2\log(r)$ as the log of a single quantity.

18. Express $3\log(4) - 1 + \log(5)$ as the log of a single number.

19. Express $10^{[3 + \log(4)]}$ as the log of a single number.

In problems 20–28, solve for x. Be careful about extraneous roots!

20. $2\log(x + 1) = \log(3x + 7)$

21. $\log(4x + 6) = \log(3) + 1$

22. $(1.07)^x = 2$ (You will have to use a calculator or a table of logs for this one.)

23. $\ln(x^2) = 16$

24. $\ln(x^2 - 5x - 5) = 0$

25. $8^x = 4^{\frac{(x+1)}{3}}$

26. $e^{2x} - 4e^x + 1 = 0$

27. $2^{(6-x)} = 4^{(2+x)}$

28. $3^{2x} - 5(3^x) + 4 = 0$

29. Prove that $a^{\log(b)} = b^{\log(a)}$.

30. Given that $\log(e) = .4343$, what is $\ln(10)$?

31. Between what two consecutive integers does $\log_2 50$ lie?

32. Solve $\log_x 36 = 2$.

Physicists and engineers often have use for what they call "hyperbolic functions" which are related to the trigonometric functions. The hyperbolic sine of x is written "$\sinh(x)$" and the hyperbolic cosine of x is written "$\cosh(x)$." These functions are defined as follows:

$$\sinh(x) = \frac{\left[e^x - e^{(-x)}\right]}{2}$$

$$\cosh(x) = \frac{\left[e^x + e^{(-x)}\right]}{2}$$

Problems 33–35 deal with hyperbolic functions.

33. Prove that $\sinh(-x) = -\sinh(x)$ and $\cosh(-x) = \cosh(x)$.

34. Prove that $\sinh(x + y) = \sinh(x)\cosh(y) + \sinh(y)\cosh(x)$ and that $\cosh(x + y) = \cosh(x)\cosh(y) + \sinh(x)\sinh(y)$.

35. What is $\sinh^2(x) + \cosh^2(x)$? What is $\sinh^2(x) - \cosh^2(x)$?

36. The loudness of sound is defined in terms of the common logarithm of the intensity level of the sound. If I represents a sound intensity that is barely audible and J is some higher level, then we say "the decibel level of J is $10 \log\left(\dfrac{J}{I}\right)$." A subway train is 10 billion times as intense as a barely audible sound. What decibel level would be given a subway train? What decibel level would be given to two subway trains?

37. It has been proved that the number of prime numbers less than n is approximately $\dfrac{n}{[\ln(n)]}$. According to this formula, how many primes would you expect less than 10,000? Note: The actual number of primes less than 10,000 is 1,229.

38. If Jane invests S dollars at the beginning of each year and each year earns $r\%$ on her total investments, then after n years she will have $\dfrac{S[(1 + r)^{(n + 1)} - 1]}{r}$. For example, if she saves $5,000 a year, earning 6% per year, then after 10 years she will have $5,000 $\dfrac{[(1.06)^{(11)} - 1]}{.06} = \$74,858.21$. Say Jane saves $5,000 a year and makes 6% annually on her investment, and her friend Simon saves $2,000 a year and makes 15% annually on his investment. Who will be the first to have total savings of $100,000? Who will be the first to have total savings of $1,000,000? How long will it take each of them to reach $1,000,000?

39. Over the period from 1926 to 1983, the geometric mean return on common stocks was 9.6% per annum. Suppose a trust had been established in 1926 with a capital of $10,000 and that each year the trust increased in value by 9.6% (before taxes) or 7.6% after taxes. How much would be in the trust in 1983? The average rate of inflation over the same period was 3% per year. What was the value of the trust in 1983, in terms of 1926 dollars?

40. The population of the United States was 75.9 million in 1900 and 226.5 million in 1980. If we assume that the population y equals a constant a times b to the nth power, where n represents the number of years and b is a constant, find the values of a and b and predict what the population will be in 2000 and in 2020.

41. If the world consumes a quantity C of some nonrenewable natural resource this year, and the consumption grows at a rate of $r\%$ per year, then over a period of n years the total consumption will be $\dfrac{C[(1 + r)^{(n+1)} - 1]}{r}$ (in this equation, for example, 6% annual growth means an r of .06). If the world consumes 17 billion barrels of oil per year, and there is a total of 2000 billion barrels of oil that can be extracted from the ground, and consumption continues to grow at 7% per year, when will we run out?

42. Radioactive substances decay according to an exponential function. If a given radioactive substance decays according to the law $N(t) = N(0)e^{(-2.3t)}$, where t is measured in years, and $N(t)$ represents the quantity of the substance remaining after t years, how many years does it take for half the substance to disappear? This period is called the *half-life* for the substance. If the law were $N(t) = N(0)e^{(-at)}$, what would be the half-life?

43. All living organisms contain a radioactive isotope of carbon called carbon 14. When the organism dies, the carbon 14 begins to decay with a half-life of 5,700 years. A mysterious man from the Caucasus approaches you with information that he has found the wreckage of Noah's Ark on Mt. Ararat. As proof he gives you a piece of cypress wood that he says was part of one of the ribs of the vessel. You measure the carbon 14 in the wood and find that there is only 80% as much as in an equivalent piece of brand-new cypress. So what do you conclude?

44. The mysterious man from the Caucasus asks your help in another project he has undertaken. He has a piece of oak that is almost exactly 1,000 years old. He wants to know in what proportions he should combine this old oak with new oak in an artifact he is building so that carbon 14 dating will indicate that the artifact is about 400 years old. What proportions do you advise?

45. Earthquakes are measured by the Richter scale. The Richter magnitude R of an earthquake is defined by the formula $R = .67\log(.37E) + 1.46$, where E is the energy in kilowatt-hours released by the quake. How many kilowatt-hours of energy were released by the San Francisco quake of 1906, which measured 7.9 on the Richter scale?

46. The Richter scale is also related to the wave amplitudes of the seismic movement of a quake. The amplitude of the seismic movement is proportional to 10^R, where R is the measure on the Richter scale. What is the ratio in the amplitudes of the seismic movement in the San Francisco quake of 1906 (7.9 on the Richter scale) to the amplitudes of the seismic movement in the San Francisco quake of 1989 (7.1 on the Richter scale)?

Check Yourself

1. $(-4)^{(-2)} = \dfrac{1}{(-4)^2} = \dfrac{1}{16}$ **(Evaluating exponentials)**

2. $5(2)^{(-2)} = \dfrac{5}{2^2} = \dfrac{5}{4}$ **(Evaluating exponentials)**

3. $\left[2^{(-1)} + 2^{(-2)}\right]^{(-1)} = \left(\dfrac{1}{2} + \dfrac{1}{4}\right)^{(-1)} = \left(\dfrac{3}{4}\right)^{(-1)} = \dfrac{4}{3}$ **(Evaluating exponentials)**

4. $[3a^{(-2)}]^3 = \left(\dfrac{3}{a^2}\right)^3 = \dfrac{27}{a^6}$ **(Evaluating exponentials)**

5. $\dfrac{8n^3 - 4n^{(-2)}}{2(n)^{(-2)}} = 4n^5 - 2$ **(Evaluating exponentials)**

6. $\dfrac{(3^5)(9^4)}{(27^4)} = \dfrac{(3^5)}{(3^4)} = 3$ **(Evaluating exponentials)**

7. $\dfrac{(b(b)^n)^2}{\left[(b^2)^n\right]} = \left(\dfrac{b^{(2n+2)}}{b^{(2n)}}\right) = b^2$ **(Evaluating exponentials)**

8. $\dfrac{2^{(-1)}}{2^{(-2)} + 2^{(-3)}} = \dfrac{\frac{1}{2}}{\left(\frac{1}{4} + \frac{1}{8}\right)} = \dfrac{\frac{1}{2}}{\frac{3}{8}} = \dfrac{4}{3}$ **(Evaluating exponentials)**

9. $\left(2^{(-1)} + 4^{(-1)}\right)^{(-1)} = \left(\dfrac{1}{2} + \dfrac{1}{4}\right)^{-1} = \left(\dfrac{3}{4}\right)^{-1} = \dfrac{4}{3}$ **(Evaluating exponentials)**

10. $\left(\dfrac{9}{25}\right)^{\frac{1}{2}} = \sqrt{\dfrac{9}{25}} = \dfrac{3}{5}$ **(Evaluating exponentials)**

11. $\log(100) = 2.\ \log(10) = 1.\ \log(1) = 0.\ \log\left(\dfrac{1}{100}\right) = -2$ **(Evaluating logarithms)**

12. $\ln 8 = \ln(2^3) = 3\ln(2) = 3(.6931) = 2.0793$ **(Evaluating logarithms)**

13. $\log(5) = \log\left(\dfrac{10}{2}\right) = \log(10) - \log(2) = 1 - .30103 = .69897$

    $\log(16) = \log(2^4) = 4\log(2) = 4(.30103) = 1.20412.\ \log(1.6) = \log\left(\dfrac{16}{10}\right) = \log(16) - 1 = .20412$ **(Evaluating logarithms)**

14. $\ln(e) = 1$; $\ln(e^3) = 3\ln(e) = 3$; $\ln(100) = \ln(10^2) = 2\ln(10) = 2(2.3026) = 4.6052$
    $\ln(27.1828) \approx \ln(10e) = \ln(10) + \ln(e) \approx 3.3026$ **(Evaluating logarithms)**

15. $\log(M) + 2\log(N) + 2 = \log(M) + \log(N^2) + \log(100) = \log(100MN^2)$ **(Evaluating logarithms)**

16. $\log(M^2N) = 2\log(M) + \log(N) = 2a + b.$ **(Evaluating logarithms)**

17. $\log(\pi) + 2\log(r) = \log(\pi r^2)$ **(Evaluating logarithms)**

18. $\log 4^3 - \log 10 + \log 5 = \log \dfrac{4^3 \cdot 5}{10} = \log \dfrac{2^6}{2^1} = \log (2^5) = \log 32$ **(Evaluating logarithms)**

19. $10^{[3+\log^{(4)}]} = \left[10^3\right]\left[10\log^{(4)}\right] = 1000(4) = 4000$ **(Evaluating logarithms)**

20. $2\log(x + 1) = \log [(x + 1)^2] = \log (3x + 7)$. So $x^2 + 2x + 1 = 3x + 7$, $x^2 - x - 6 = 0$, and $(x - 3)(x + 2) = 0$.
    So the possible roots are $x = 3$ and $x = -2$. Since $\log(-1)$ is not defined, the root $x = -2$ is extraneous,
    but $x = 3$ is a root. **(Solving exponential and logarithmic equations)**

21. $\log(4x + 6) = \log(3) + 1 = \log(30)$. So $4x + 6 = 30$, $4x = 24$, and $x = 6$. **(Solving exponential and logarithmic equations)**

22. $(1.07)^x = 2$, so $x\log(1.07) = \log(2) = .30103$, $x = \dfrac{.30103}{\log(1.07)} = 10.24477$ **(Solving exponential and logarithmic equations)**

23. $16 = \ln(x^2) = 2\ln(x)$, so $\ln(x) = 8$, so $x = e^8$. **(Solving exponential and logarithmic equations)**

24. $\ln(x^2 - 5x - 5) = 0 = \ln(1)$, so $x^2 - 5x - 5 = 1$, $x^2 - 5x - 6 = 0$, and $(x - 6)(x + 1) = 0$. So possible roots are
    $x = 6$ and $\mathrm{x} = -1$. Both these roots solve the equation. **(Solving exponential and logarithmic equations)**

25. $x\log(8) = \left[\dfrac{(x + 1)}{3}\right]\log (4)$. $3x\log(2) = \left[\dfrac{2(x + 1)}{3}\right]\log(2)$. $3x = \dfrac{2x + 2}{3}$. $7x = 2$. $x = \dfrac{2}{7}$ **(Solving exponential and logarithmic equations)**

26. Let $u = e^x$. $u^2 - 4u + 1 = 0$. $u = \left[\dfrac{4 + \sqrt{12}}{2}\right]$ or $u = \left[\dfrac{4 - \sqrt{12}}{2}\right]$. So $x = \ln(2+\sqrt{3})$ or $x = \ln(2-\sqrt{3})$. **(Solving exponential and logarithmic equations)**

27. $2^{(6 - x)} = 2^{[2(2+x)]}$, so $6 - x = 2(2 + x)$, $2 = 3x$, $x = \dfrac{2}{3}$. **(Solving exponential and logarithmic equations)**

28. Let $3^x = u$. Then $u^2 - 5u + 4 = 0$, $(u - 1)(u - 4) = 0$, so $3^x = 1$ or $3^x = 4$, so $x = 0$ or $x = \dfrac{\log(4)}{\log(3)}$. **(Solving exponential and logarithmic equations)**

29. $\log(b)\log(a) = \log(a)\log(b)$ by the commutative law of multiplication, therefore
    $\log(a^{[\log(b)]}) = \log(b^{[\log(a)]})$, and therefore $a^{\log(b)} = b^{\log(a)}$. **(Solving exponential and logarithmic equations)**

30. By the change of base formula, $\ln(10) = \log_e 10 = \dfrac{\log(10)}{\log(e)} = \dfrac{1}{.4343} = 2.3026.$ (**Solving exponential and logarithmic equations**)

31. Since $2^5 < 50 < 2^6$, it follows that $\log_2(50)$ is between 5 and 6. (**Solving exponential and logarithmic equations**)

32. $x^2 = 36$, therefore $x = 6$ or $x = -6$. Logarithms to negative bases are not defined, so $x = 6$ is the sole solution. (**Solving exponential and logarithmic equations**)

33. $\sinh(-x) = \dfrac{\left[e^{(-x)} - e^{[-(-x)]}\right]}{2} = \dfrac{\left[e^{(-x)} - e^{(x)}\right]}{2} = -\sinh(x).$ $\cosh(-x) = \dfrac{\left[e^{(-x)} + e^{[-(-x)]}\right]}{2} = \dfrac{\left[e^{x} + e^{(-x)}\right]}{2} = \cosh(x).$
(**Solving exponential and logarithmic equations**)

34. It is best to start with the right side of the expression, so take
$$\sinh(x)\cosh(y) + \sinh(y)\cosh(x) = \dfrac{\left[(e^x - e^{(-x)})(e^y + e^{(-y)}) + (e^y - e^{(-y)})(e^x + e^{(-x)})\right]}{4} =$$
$$\dfrac{\left[e^{(x+y)} + e^{(x-y)} - e^{(y-x)} - e^{(-x-y)} + e^{(x+y)} + e^{(y-x)} - e^{(x-y)} - e^{(-x-y)}\right]}{4} = 2\dfrac{\left[e^{(x+y)} - e^{(-x-y)}\right]}{2} = \sinh(x+y).$$ To
prove $\cosh(x + y) = \cosh(x)\cosh(y) + \sinh(x)\sinh(y)$, start with the right-hand side, expand and simplify, and the result follows. (**Solving exponential and logarithmic equations**)

35. $\sinh^2(x) + \cosh^2(x) = \dfrac{\left[e^{2x} - 2e^x e^{(-x)} + e^{(-2x)} + e^{2x} + 2e^x e^{(-x)} + e^{(-2x)}\right]}{4} = \dfrac{\left[e^{2x} + e^{(-2x)}\right]}{2} = \cosh(2x).$ Note that
$\sinh(x) + \cosh(x) = e^x$, and $\sinh(x) - \cosh(x) = e^{(-x)}$, so
$\sinh^2(x) - \cosh^2(x) = [\sinh(x) + \cosh(x)][\sinh(x) - \cosh(x)] = e^x e^{(-x)} = 1.$ (**Solving exponential and logarithmic equations**)

36. Ten billion is 10^{10}, so $10\log\left(\dfrac{10^{10}}{1}\right) = 100$ decibels is the level of one subway train. The intensity of two subway trains would be $2(1,010)$, and we have $10\log[2(10^{10})] = 10[10 + \log(2)] = 10[10.30103] \approx 103$ decibels. (**Applications of exponents and logarithms**)

37. $\ln(10,000) = 4\ln(10) = 4(2.3026) \approx 9.21$, so $\dfrac{10,000}{\ln(10,000)} \approx 1086$, and so the formula predicts about 1,086 primes less than 10,000. (**Applications of exponents and logarithms**)

38. Jane, who saves \$5,000 annually, will break \$100,000 after 13 years and will break \$1,000,000 after 44 years. Her friend Simon, who saves \$2,000 annually, will break \$100,000 after 15 years and will break \$1,000,000 after 30 years. After 44 years, he will have about \$7,100,000. (**Applications of exponents and logarithms**)

39. If paying taxes could somehow be avoided, the trust would be worth $(1.096)^{57}$ times \$10,000 in 1983, or \$1,858,667. But since taxes must be paid the trust will be worth only $(1.076)^{57}$ times \$10,000, or \$650,578. To convert to 1926 dollars, divide by $(1.03)^{57} \approx 5.39$, which deflates the value of the trust to \$120,664 in 1926 dollars. (**Applications of exponents and logarithms**)

40. Taking 1900 as year 0 we get $a = 75.9$. Then 1980 corresponds to year 80, and we have

$226.5 = (75.9)(b^{80})$, so $b^{80} = 2.9842$, and $\log(b) = \left(\dfrac{1}{80}\right)\log(2.9842) = .0059353$, so $b = 1.01376$. The population will be 297.7 million in 2000 and 391.3 million in 2020. (**Applications of exponents and logarithms**)

41. Simply solve $\dfrac{17\left[(1.07)^{(n+1)} - 1\right]}{(.07)} = 2000$ for n, and you find that if this rate continues, we will run out of oil in about 32 years. (**Applications of exponents and logarithms**)

42. Solve $e^{(-2.3t)} = \dfrac{1}{2}$ for t (take the natural logarithm of both sides) and you get a half-life of about .3 years.

If $N(t) = N(0)e^{(-at)}$, then the half-life is $\dfrac{\ln\left(\dfrac{1}{2}\right)}{-a} = \dfrac{\ln(2)}{a}$. (**Applications of exponents and logarithms**)

43. With a half-life of 5,700 years, carbon 14 obeys the law $N(t) = N(0)e^{(-at)}$ with $a = \dfrac{\ln(2)}{5700} = .0001216$. If

$e^{(-at)} = .8$, it follows that $t = \dfrac{-\ln(.8)}{(.0001216)} \approx 1835$ years. Since the ark is supposed to have been built before the time of Christ, it is impossible that this piece of wood came from Noah's Ark. (**Applications of exponents and logarithms**)

44. A 400-year-old piece of oak would have a carbon 14 concentration of $e^{(-a400)}$, or .9525. The 1,000-year-old piece of oak has a carbon 14 concentration of $e^{(-a1000)}$, or .8855. New oak has a concentration of 1.0000 carbon 14. A mixture of 41% of the old oak and 59% of the new oak would have the same overall carbon 14 content as a piece of oak 400 years old. But how can the man combine these two pieces of oak so that a suspicious curator could not take a sample of a small piece of the wood and thereby discover that the little piece he selects is either 1,000 years old or brand-new, in either case undermining the case for the genuineness of the artifact? (**Applications of exponents and logarithms**)

45. Solve the equation $7.9 = .67\log(.37E) + 1.46$ for E. You should get about 11 billion kilowatt-hours. (**Applications of exponents and logarithms**)

46. $\dfrac{\left(10^{7.9}\right)}{\left(10^{7.1}\right)} = 10(7.9 - 7.1) = 10(.8) = 6.31$, so the waves had amplitude approximately 6.3 times as great in

1906 compared to the waves in 1989. (**Applications of exponents and logarithms**)

Grade Yourself

Circle the numbers of the questions you missed, then fill in the total incorrect for each topic. If you answered more than three questions incorrectly, you need to focus on that topic. (If a topic has fewer than three questions and you had at least one wrong, we suggest you study that topic also. Read your textbook or a review book, or ask your teacher for help.)

Subject: Exponents and Logarithms

Topic	Question Numbers	Number Incorrect
Evaluating exponentials	1, 2, 3, 4, 5, 6, 7, 8, 9, 10	
Evaluating logarithms	11, 12, 13, 14, 15, 16, 17, 18, 19	
Solving exponential and logarithmic equations	20, 21, 22, 23, 24, 25, 26, 27, 28, 29, 30, 31, 32, 33, 34, 35	
Applications of exponents and logarithms	36, 37, 38, 39, 40, 41, 42, 43, 44, 45, 46	

Trigonometry: Basic Definitions and Solving Triangles

5

Brief Yourself

Measuring Angles

Two different units are commonly used for measuring angles. In surveying, navigation, and geometry, degrees are ordinarily used. *Degree measure* divides the arc of the circle into 360 equal parts. In engineering, physics, and higher mathematics, radians are ordinarily used. *Radian measure* divides the arc of the circle into 2π parts. If an angle is placed at the center of a circle, the radian measure of the angle will be the ratio of the length of the arc it subtends to the radius of the circle. If D is the measure of an angle in degrees and R is the measure of the same angle in radians, then

$$\frac{D}{180} = \frac{R}{\pi}$$

Defining Basic Trigonometric Functions

Consider a right triangle. Let the angles in the triangle be A, B, and $C = 90°$. Note that A and B are complements of one another; that is to say, they add to $90°$. Let a, b, and c be the lengths of the sides opposite these angles. Then the trigonometric functions of these angles are defined as follows:

$$\tan(A) = \frac{a}{b} \qquad \tan(B) = \frac{b}{a}$$

$$\sin(A) = \frac{a}{c} \qquad \sin(B) = \frac{b}{c}$$

$$\cos(A) = \frac{b}{c} \qquad \cos(B) = \frac{a}{c}$$

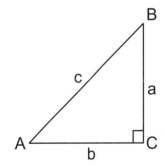

$$\sec(A) = \frac{c}{b} \qquad\qquad \sec(B) = \frac{c}{a}$$

$$\csc(A) = \frac{c}{a} \qquad\qquad \csc(B) = \frac{c}{b}$$

$$\cot(A) = \frac{b}{a} \qquad\qquad \cot(B) = \frac{a}{b}$$

These six functions—the sine and cosine, tangent and cotangent, secant and cosecant—are really only three different functions, for the cosine of an angle is the sine of the angle's complement, the cotangent of an angle is the tangent of the angle's complement, and the cosecant is the secant of the angle's complement. It is also easy to see from these definitions that the functions are closely related to one another. For example, it immediately follows that

$$\tan(A) = \frac{\sin(A)}{\cos(A)}$$

$$\cot(A) = \frac{\cos(A)}{\sin(A)} = \frac{1}{\tan(A)}$$

$$\sec(A) = \frac{1}{\cos(A)}$$

$$\csc(A) = \frac{1}{\sin(A)}$$

In most cases it is necessary to use a table or a calculator to get the value of a trigonometric function of an angle. But there are a few angles for which the trigonometric functions may be easily calculated in terms of square roots or rational numbers. Here is a brief list of some of these special angles for the sine and cosine functions:

$$\sin(0°) = 0 \qquad\qquad \cos(0°) = 1$$

$$\sin(30°) = \frac{1}{2} \qquad\qquad \cos(30°) = \frac{\sqrt{3}}{2}$$

$$\sin(45°) = \frac{1}{\sqrt{2}} \qquad\qquad \cos(45°) = \frac{1}{\sqrt{2}}$$

$$\sin(60°) = \frac{\sqrt{3}}{2} \qquad\qquad \cos(60°) = \frac{1}{2}$$

$$\sin(90°) = 1 \qquad\qquad \cos(90°) = 0$$

With each of the trigonometric functions, there is associated an inverse trigonometric function. For example, for the sine function we have the arcsine function. If A is an acute angle, and $\sin(A) = x$, then $\arcsin(x) = A$. The arccos, arctan, arcsec, arccot, and arccsc are defined in an analogous way.

Defining trigonometric functions in terms of the angles and sides of a right triangle is the traditional way to define the trigonometric functions, but for many applications it is more convenient to define them in terms of the *unit circle*. Draw a circle of radius 1 about the origin in an *x-y* coordinate system. Consider a ray from the origin (0, 0) that makes an angle θ with the *x*-axis. This ray will cut the unit circle at a point (*x*, *y*). It is easy to see that if θ is an acute angle then $x = \cos\theta$ and $y = \sin\theta$. We simply extend this around the circle and define the values of cosθ and sinθ to be the *y* and *x* coordinates of the point where the ray intersects the unit circle. The *x*- and *y*-axes divide the Cartesian plane into four quarters. The quarter containing acute angles θ is called the *first quadrant*, that containing values of θ between 90° and 180° is called the *second quadrant*, that containing values of θ between 180° and 270° is called the *third quadrant*, and that containing values of θ between 270° and 360° is called the *fourth quadrant*.

Values of the trigonometric functions tangent, cotangent, secant, and cosecant can be defined in terms of the sine and cosine functions when you take the unit circle approach.

Solving Triangles

It is easy to see how trigonometric functions can be used to find unknown sides and unknown angles in right triangles. These functions can, in fact, be used to find the unknown parts of many triangles by using two general formulas. The first of these is the law of sines. If *A*, *B*, *C* are the angles of any triangle (not necessarily a right triangle) and *a*, *b*, *c* are the lengths of the sides opposite these angles, then

$$\frac{\sin(A)}{a} = \frac{\sin(B)}{b} = \frac{\sin(C)}{c}$$

The second of these formulas is the law of cosines:

$$c^2 = a^2 + b^2 - 2ab\cos(C)$$

$$b^2 = a^2 + c^2 - 2ac\cos(B)$$

$$a^2 = b^2 + c^2 - 2bc\cos(A)$$

Note that the law of cosines turns into the Pythagorean theorem if the angle is a right angle, and the law of sines is a direct consequence of the definition of the sine function if one of the angles is a right angle.

Test Yourself

1. What is the measure of a right angle in radians? In degrees?

2. If two of the interior angles of a triangle are complementary to one another, what can you say about the third angle in the triangle?

3. If one of the base angles of a right triangle is 30°, what is the measure of the other base angle?

4. What is the measure of a 30° angle in radians? What is the measure of a 60° angle in radians?

5. How many radians are in one degree?

6. How many degrees are in one radian?

7. Convert each of the following angles to radian measure:

   (a) 10°

   (b) 45°

   (c) 75°

   (d) 135°

   (e) 270°

8. Convert each of the following angles to degree measure:

   (a) 2π radians

   (b) π radians

   (c) $\frac{\pi}{6}$ radians

   (d) $\frac{3\pi}{2}$ radians

   (e) 1.2 radians

For problems 9–18, you should not require the use of tables or a calculator.

9. If the sides of a right triangle are 3, 4, and 5 units, what are the sine and the cosine of the angle opposite the side of length 4 units?

10. If the sides of a right triangle are 5, 12, and 13 units, what are the sine and the cosine of the angle opposite the side of length 12 units?

11. Suppose you have a triangle ABC in which $C = 90°$ and $\sin(A) = \frac{3}{5}$. If the longest side of the triangle is 10 feet, what are the lengths of the other two sides of the triangle? What is the $\sin(B)$?

12. If a ladder 12 feet long is leaning against a wall and makes an angle of 60° with the ground, how far up the wall does it reach?

13. Suppose you have a triangle with angles 30°, 60°, and 90°. If the shortest side has length of 5 meters, what are the lengths of the other two sides?

14. A baseball diamond is a perfect square, 90 feet along each side. How far is it from home plate to second base?

15. What is the area of an equilateral triangle 2 units on a side? (Hint: The area of a triangle is one-half the base times the height).

16. If A is an acute angle with $\sin(A)$ greater than $\cos(A)$, what can you conclude about the size of angle A?

17. If one side of a rectangular field is 50 feet and the diagonal distance from one corner to the opposite corner is 130 feet, what is the length of the other side of the field? What is the sine of the angle between the short side of the field and a diagonal? What is the cosine of the same angle?

18. If an SR-71 observation airplane is directly over the head of an observer at point x and the line of sight from an observer at point y, 50,000 feet from point x, makes an angle of 60° with the ground, then what is the straight-line distance of the plane from the observer at point y? How high is the plane above the ground?

19. Express sin(170°) as the sine of an acute angle.

20. Express sin(100°) as the sine of an acute angle.

21. Express sin(190°) as the sine of an acute angle.

22. Express sin(225°) as the sine of an acute angle.

23. Express sin(300°) as the sine of an acute angle.

24. Express cos(150°) as the cosine of an acute angle.

25. Express cos(120°) as the cosine of an acute angle.

26. Express cos(210°) as the cosine of an acute angle.

27. Express cos(240°) as the cosine of an acute angle.

28. Express cos(315°) as the cosine of an acute angle.

29. For what values of β are cos(β) and sin(β) both positive?

30. For what values of β are cos(β) positive and sin(β) negative?

31. For what values of β are cos(β) and sin(β) both negative?

32. For what values of β are cos(β) negative and sin(β) positive?

33. Find the exact value of the sine and cosine of 135°.

34. Find the exact value of the sine and cosine of 210°.

35. Find the exact value of the sine and cosine of 180°.

36. Find the exact value of the sine and cosine of 150°.

37. Find the exact value of the sine and cosine of 315°.

38. Find the exact value of the sine and cosine of 300°.

39. Find the exact value of the sine and cosine of 390°.

40. Find the exact value of the sine and cosine of 720°.

41. Find the exact value of the sine and cosine of 1000°.

42. Find the exact value of the sine and cosine of −30°.

43. Find the exact value of the sine and cosine of −420°.

In problems 44–53, find the other angles and the other sides for the triangle with angles A, B, and C and sides opposite these angles of length a, b, and c. If there is more than one solution, find both.

44. $A = 45°, B = 45°, c = 20$

45. $A = 75°, B = 25°, c = 30$

46. $A = 100°, B = 20°, a = 10$

47. $A = 60°, b = 20, c = 20$

48. $A = 30°, a = 10, b = 19$

49. $A = 30°, a = 10, b = 20$

50. $A = 30°, a = 10, b = 21$

51. $a = 3, b = 4, c = 5$

52. $a = 1, b = \dfrac{1 + \sqrt{5}}{2}, c = \dfrac{1 + \sqrt{5}}{2}$

53. $a = 5, b = 4, c = 7$

Check Yourself

1. A right angle is one-fourth of a full circle. A full circle is 2π radians, and so a right angle is $\frac{\pi}{2}$ radians. The same argument shows that a right angle is 90°. (**Measuring angles**)

2. The interior angles of a triangle always add up to 180° (or π radians). If two of the angles are complementary, they add up to 90° (or $\frac{\pi}{2}$ radians). Therefore, the third angle must be 90° (or $\frac{\pi}{2}$ radians) and the triangle is a right triangle. (**Measuring angles**)

3. Since the base angles of a right triangle are complementary to one another, if one base angle is 30°, the other must be 60°. (**Measuring angles**)

4. A 30° angle is one-sixth of a straight angle. One-sixth of a straight angle is $\frac{\pi}{6}$ radians. A 60° angle is one-third of a straight angle. One-third of a straight angle is $\frac{\pi}{3}$ radians. Or you could perform the calculation $\frac{30}{180} = \frac{\beta}{\pi}$ and find $\beta = \frac{\pi}{6}$. Similarly, $\frac{60}{180} = \frac{\beta}{\pi}$ gives $\beta = \frac{\pi}{3}$. (**Measuring angles**)

5. Use the formula $\frac{1}{180} = \frac{\beta}{\pi}$ and you get $\beta = \frac{\pi}{180} = \frac{3.14159...}{180} = .01745$. (**Measuring angles**)

6. Use the formula $\frac{\pi}{180} = \frac{1}{\beta}$ and you get $\beta = \frac{180}{\pi} = \frac{180}{3.14159...} = 57.2958...°$. (**Measuring angles**)

7. Simply use the formula. (a) $\frac{10}{180} = \frac{\beta}{\pi}$ which gives $\beta = \frac{\pi}{18} = .1745...$ (b) $\frac{45}{180} = \frac{\beta}{\pi}$, which gives $\beta = \frac{\pi}{4} = .7854...$ (c) $\frac{75}{180} = \frac{\beta}{\pi}$, which gives $\beta = \frac{5\pi}{12} = 1.309...$ (d) $\frac{135}{180} = \frac{\beta}{\pi}$, which gives $\beta = \frac{3\pi}{4} = 2.3562...$ (e) $\frac{270}{180} = \frac{\beta}{\pi}$, which gives $\beta = \frac{3\pi}{2} = 4.7124...$ (**Measuring angles**)

8. (a) Since 2π is a complete circle, an angle with radian measure 2π has degree measure 360°. (b) Therefore, an angle with radian measure π has degree measure 180°.

   For the other three, use the formula

   (c) $\frac{\pi}{6}$ radians: $\frac{\beta}{180} = \frac{\frac{\pi}{6}}{\pi}$, which gives $B = \frac{180}{6} = 30°$.

   (d) $\frac{3\pi}{2}$ radians: $\frac{B}{180} = \frac{\frac{3\pi}{2}}{\pi}$, which gives $B = \frac{540}{2} = 270°$.

   (e) 1.2 radians: $\frac{B}{180} = \frac{1.2}{\pi}$, which gives $B = \frac{216}{\pi} = 68.7549...°$. (**Measuring angles**)

9. The sine of the angle opposite the side of length 4 units will be $\frac{4}{5}$, and its cosine will be $\frac{3}{5}$. (**Basic definitions**)

10. The sine of the angle opposite the side of length 12 units will be $\frac{12}{13}$, and its cosine will be $\frac{5}{13}$. (**Basic definitions**)

11. The shortest side will be 6 feet, and the third side will be 8 feet. You will have $\sin(B) = \frac{8}{10} = \frac{4}{5}$. (**Basic definitions**)

12. $12\sin(60) = 12\frac{\sqrt{3}}{2} = 6\sqrt{3} = 10.4$ feet (approximately). (**Solving triangles**)

13. The longest side will have length of 10 meters, and the third side will have length of 8.7 meters (approximately). (**Solving triangles**)

14. $90\sqrt{2} = 127.3$ feet (approximately). (**Solving triangles**)

15. The altitude will be $\sqrt{3}$, so the total area is $2\frac{\sqrt{3}}{2} = \sqrt{3}$. (**Solving triangles**)

16. If the sine of an acute angle is greater than the sine of its complement, then the angle itself must be greater than its complement. So it follows that A must be greater than 45°. (**Basic definitions**)

17. By the Pythagorean theorem, the other side of the field must be 120 feet. The sine of the angle between the short side and the diagonal must be $\frac{12}{13}$, and the cosine of the same angle must be $\frac{5}{13}$. (**Basic definitions**)

18. If d is the straight-line distance from observer y, then $d = \frac{50,000}{\cos 60°}$, $\cos(60°) = \frac{1}{2}$ so the plane is 100,000 feet from point y. $100,000\sin(60°) = 86,600$ feet altitude (approximately). You can see why we specified that the plane was an SR-71. (**Solving triangles**)

19. $\sin(170°) = \sin(10°)$ (**Basic definitions**)

20. $\sin(100°) = \sin(80°)$ (**Basic definitions**)

21. $\sin(190°) = -\sin(10°)$ (**Basic definitions**)

22. $\sin(225°) = -\sin(45°)$ (**Basic definitions**)

23. $\sin(300°) = -\sin(60°)$ (**Basic definitions**)

24. $\cos(150°) = -\cos(30°)$ (**Basic definitions**)

25. $\cos(120°) = -\cos(60°)$ (**Basic definitions**)

26. $\cos(210°) = -\cos(30°)$ (**Basic definitions**)

27. $\cos(240°) = -\cos(60°)$ **(Basic definitions)**

28. $\cos(315°) = \cos(45°)$ **(Basic definitions)**

29. If $0° < \beta < 90°$, then sine and cosine are both positive. Also, if for any integer n we have $0° < \beta + 360n < 90°$, then sine and cosine of β are both positive. **(Basic definitions)**

30. If $270° < \beta < 360°$, then sine is negative and cosine is positive. Also, if for any integer n we have $270° < \beta + 360n < 360°$, then sine is negative and cosine is positive. **(Basic definitions)**

31. If $180° < \beta < 270°$, then sine and cosine are both negative. Also, if for any integer n we have $180° < \beta + 360n < 270°$, then sine and cosine of β are both negative. **(Basic definitions)**

32. If $90° < \beta < 180°$, then sine is positive and cosine is negative. Also, if for any integer n we have $90° < \beta + 360n < 180°$, then sine is positive and cosine is negative. **(Basic definitions)**

33. $\sin(135°) = \dfrac{\sqrt{2}}{2}; \quad \cos(135°) = \dfrac{-\sqrt{2}}{2}$ **(Basic definitions)**

34. $\sin(210°) = \dfrac{-1}{2}; \cos(210°) = \dfrac{-\sqrt{3}}{2}$ **(Basic definitions)**

35. $\sin(180°) = 0; \cos(180°) = -1$ **(Basic definitions)**

36. $\sin(150°) = \dfrac{1}{2}; \cos(150°) = \dfrac{-\sqrt{3}}{2}$ **(Basic definitions)**

37. $\sin(315°) = \dfrac{-\sqrt{2}}{2}; \cos(315°) = \dfrac{\sqrt{2}}{2}$ **(Basic definitions)**

38. $\sin(300°) = \dfrac{-\sqrt{3}}{2}; \cos(300°) = \dfrac{1}{2}$ **(Basic definitions)**

39. $\sin(390°) = \sin(30°) = \dfrac{1}{2}; \cos(390°) = \cos(30°) = \dfrac{\sqrt{3}}{2}$ **(Basic definitions)**

40. $\sin(720°) = \sin(0°) = 0; \cos(720°) = \cos(0°) = 1$ **(Basic definitions)**

41. $\sin(1000°) = \sin(280°) = -\sin(80°) = -.9848; \cos(1000°) = \cos(280°) = \cos(80°) = .1736$ **(Basic definitions)**

42. $\sin(-30°) = -\sin(30°) = \dfrac{1}{2}; \cos(-30°) = \cos(30°) = \dfrac{\sqrt{3}}{2}$ **(Basic definitions)**

43. $\sin(-420°) = -\sin(420°) = -\sin(60°) = \dfrac{\sqrt{3}}{2}; \cos(-420°) = \cos(420°) = \cos(60°) = \dfrac{1}{2}$ **(Basic definitions)**

44. $C = 180° - 45° - 45° = 90°; a = b = 20\sin(45°) = 14.142$ **(Solving triangles)**

45. $C = 180° - 75° - 25° = 80°; \dfrac{\sin 80°}{30} = \dfrac{\sin 75°}{a}$, so $a = \dfrac{30\sin 75°}{\sin 80°} = 29.425$, and $\dfrac{30\sin 25°}{\sin 80°} = 12.874$ **(Solving triangles)**

46. $C = 180° - 100° - 20° = 60°$; $\dfrac{\sin 60°}{c} = \dfrac{\sin 100°}{10}$, so $c = \dfrac{10\sin 60°}{\sin 100°} = 8.794$, and $b = \dfrac{10\sin 20°}{\sin 100°} = 3.473$ (**Solving triangles**)

47. By the law of cosines, $a^2 = 20^2 + 20^2 - 2(20^2)\cos(60°) = 400$, so $a = 20$ and we have an equilateral triangle, which means $B = C = 60°$. (**Solving triangles**)

48. There are two solutions, because $\dfrac{19\sin 30°}{10} = \dfrac{19}{20} < 1$. $\text{Sin}(B) = \dfrac{19}{20} = .95$ implies that $B = 71.8°$ or $B = 108.2°$. In the former case, we have $A = 30°$, $B = 71.8°$, $C = 78.2°$ and $a = 10$, $b = 19$, $c = 19.58$. In the latter case, we have $A = 30°$, $B = 108.2°$, $C = 41.8°$ and $a = 10$, $b = 19$, $c = 13.33$. (**Solving triangles**)

49. There is just one solution, because $\dfrac{20\sin 30°}{10} = \dfrac{20}{20} = 1$. $B = 90°$, so $C = 60°$. The sides are $a = 10$, $b = 20$, $c = 17.32$. Note that in the ambiguous case, if there is just one solution, one of the angles must be $90°$. (**Solving triangles**)

50. There is no solution, because $\dfrac{21\sin 30}{10} = \dfrac{21}{20} > 1$, so there is no possible value for angle B. (**Solving triangles**)

51. By the law of cosines, $25 = 16 + 9 - 2(3)(4)\cos(C)$, so $\cos(C) = 0$, which means $C = 90°$. We can apply the law of cosines to side b and get $16 = 25 + 9 - 2(5)(3)\cos(B)$ so $-18 = -30\cos(B)$, which gives $\cos(B°) = .6$ so $B = 53.13°$ and therefore $A = 36.87°$. (**Solving triangles**)

52. Apply the law of cosines to angle B, using the fact that $c = b$, and you have $b^2 = b^2 + 1 - 2b\cos(B)$, so $\cos(B) = \dfrac{1}{1 + \sqrt{5}}$, which gives $B = 72°$ exactly. Therefore, $C = 72°$ and $A = 36°$. (**Solving triangles**)

53. Apply the law of cosines to angle B and you get $\cos(B) = \dfrac{58}{70}$, so $B = 34.05°$. Apply the law of cosines to angle A and you get $\cos(A) = \dfrac{40}{56}$, so $A = 44.42°$. Apply the law of cosines to angle C and you get $\cos(C) = \dfrac{-8}{40}$, so $C = 101.53°$. The fact that these angles add up to $180°$ verifies that we haven't made any numerical errors. (**Solving triangles**)

Grade Yourself

Circle the numbers of the questions you missed, then fill in the total incorrect for each topic. If you answered more than three questions incorrectly, you need to focus on that topic. (If a topic has fewer than three questions and you had at least one wrong, we suggest you study that topic also. Read your textbook or a review book, or ask your teacher for help.)

Subject: Trigonometry: Basic Definitions and Solving Triangles

Topic	Question Numbers	Number Incorrect
Measuring angles	1, 2, 3, 4, 5, 6, 7, 8	
Basic definitions	9, 10, 11, 16, 17, 19, 20, 21, 22, 23, 24, 25, 26, 27, 28, 29, 30, 31, 32, 33, 34, 35, 36, 37, 38, 39, 40, 41, 42, 43	
Solving triangles	12, 13, 14, 15, 18, 44, 45, 46, 47, 48, 49, 50, 51, 52, 53	

Trigonometry: Identities, Equations, and Applications

6

Brief Yourself

What Are Trigonometric Identities?

Trigonometric identities are trigonometric expressions that are true for all values of the variables for which the expressions are defined. The basic trigonometric identities fall into three categories: *fundamental identities*, *Pythagorean identities*, and *addition formulas*. From these basic identities other identities can be constructed. Proving and creating trigonometric identities is a time-proven way of learning to work with trigonometric functions, and any test of trigonometric knowledge is certain to include some questions about identities.

Fundamental Identities

Fundamental identities are identities that are immediate consequences of the definitions of the trigonometric functions:

$$\tan(x) = \frac{\sin(x)}{\cos(x)} \qquad \cot(x) = \frac{\cos(x)}{\sin(x)} \qquad \cot(x) = \frac{1}{\tan(x)}$$

$$\sec(x) = \frac{1}{\cos(x)} \qquad \csc(x) = \frac{1}{\sin(x)} \qquad \tan(x) = \frac{\sec(x)}{\csc(x)}$$

$$\cot(x) = \frac{\csc(x)}{\sec(x)} \qquad \cos(x) = \sin(90° - x) \qquad \csc(x) = \sec(90° - x)$$

$$\cot(x) = \tan(90° - x)$$

Pythagorean Identities

Three Pythagorean identities are a direct consequence of the Pythagorean theorem.

$$\cos^2(x) + \sin^2(x) = 1$$

$$1 + \tan^2(x) = \sec^2(x)$$

$$1 + \cot^2(x) = \csc^2(x)$$

Addition Formulas

There are many identities that involve the values of trigonometric functions of angle sums and angle differences. Here is a list of some of the most common ones:

$$\sin(x + y) = \sin(x)\cos(y) + \sin(y)\cos(x)$$ $$\sin(x - y) = \sin(x)\cos(y) - \sin(y)\cos(x)$$

$$\cos(x + y) = \cos(x)\cos(y) - \sin(x)\sin(y)$$ $$\cos(x - y) = \cos(x)\cos(y) + \sin(x)\sin(y)$$

$$\tan(x + y) = \frac{\tan(x) + \tan(y)}{1 - \tan(x)\tan(y)}$$ $$\tan(x - y) = \frac{\tan(x) - \tan(y)}{1 + \tan(x)\tan(y)}$$

$$\cot(x + y) = \frac{\cot(x)\cot(y) - 1}{\cot(x) + \cot(y)}$$ $$\cot(x - y) = \frac{\cot(x)\cot(y) + 1}{\cot(x) - \cot(y)}$$

$$\sin(x+y)\sin(x-y) = \sin^2(x) - \sin^2(y) = \cos^2(y) - \cos^2(x)$$

$$\cos(x+y)\cos(x-y) = \cos^2(x) - \sin^2(y) = \cos^2(y) - \sin^2(x)$$

From these identities, it is easy to derive formulas for $\sin(2x)$, $\cos(2x)$, $\tan(2x)$, $\sin(3x)$, and so on. For example, $\sin(2x) = \sin(x + x) = \sin x \cos x + \sin x \cos x = 2\sin x \cos x$. Similarly, $\cos(2x)$ can be shown to be equal to $\cos^2(x) - \sin^2(x)$.

What Are Trigonometric Equations?

A trigonometric equation is an expression that is true only for some values of the variables. It differs from an identity in that an identity is true for all values of the variables for which the expression is defined. This distinction between equations and identities is also found in algebra. For example, $x^2 - y^2 = (x + y)(x - y)$ is an algebraic identity, true for all values of x and y. But $x^2 - 5x - 6 = 0$ is an algebraic equation, true only for $x = 6$ and $x = -1$.

An example of an equation from trigonometry is $\cos^2(x) + \sin^2(x) = 1$.

From the addition formula for the cosine above, we see that this is equivalent to $\cos(2x) = 1$.

This equation is true only for $2x = 0, 360°$. This can be written $k360°$ for all integer values of k. So the solution to the trigonometric equation is $x = k180°$.

Solving trigonometric equations often requires the use of a variety of algebraic and trigonometric tools.

Applications to Complex Numbers

One of the most important applications of trigonometric functions is to the area of complex analysis. Using i to stand for the square root of -1, we can identify complex numbers $z = x + iy$ with points in the (x, y) plane. Introducing polar coordinates into the plane gives us the expression

$z = r[\cos(\beta) + i\sin(\beta)]$ where $r = x^2 + y^2$ and $\beta = \arctan\left(\dfrac{y}{x}\right)$ (choosing β in the first quadrant if x and y

are both positive, in the second quadrant if x is negative and y is positive, in the third quadrant if both x and y are negative, and in the fourth quadrant if x is positive and y is negative.)

The notation "$r\text{cis}(x)$" is often used for "$r[\cos(x) + i\sin(x)]$" in order to save space. It follows immediately from the laws for multiplying complex numbers that if

$w = r[\cos(a) + i\sin(a)]$ and $z = s[\cos(b) + i\sin(b)]$ then
$wz = rs[\cos(a)\cos(b) - \sin(a)\sin(b) + i\sin(a)\cos(b) + i\cos(a)\sin(b)] = rs[\cos(a + b) + i\sin(a + b)]$.

From this we derive De Moivre's Formula for finding powers and roots of complex numbers. Specifically, if $z = r[\cos(\beta) + i\sin(\beta)]$, then

$$z^n = r^n\,[\cos(n\beta) + i\sin(n\beta)]$$

The rule applies for any rational value of n. If n has a denominator p, there will be p multiple roots. These can be found by taking β, $\beta + 360°$, and so on up to $\beta + (n - 1)360°$ and applying the formula.

De Moivre's Formula gives us a convenient way to find formulas for $\sin(3x)$, $\cos(3x)$, $\sin(4x)$, $\cos(4x)$, etc., in terms of $\sin(x)$ and $\cos(x)$. We simply take the appropriate power of $[\cos(x) + i\sin(x)]$ in two different ways (that is, use De Moivre's Theorem and use the Binomial theorem) and equate the real and imaginary parts of the two results.

Polar Coordinates

Another way of associating the points in a plane with pairs of numbers is through polar coordinates. If you have a plane that already has a Cartesian coordinate system, you can impose a polar coordinate system by means of the transformation:

$$r = \sqrt{(x^2 + y^2)}$$

$$\theta = \arctan\left(\frac{y}{x}\right)$$

r denotes the distance of the point P from the origin $(0, 0)$, and θ represents the angle that the line segment from $(0, 0)$ through (x, y) makes with the x-axis. You must be careful to choose θ so that if x and y are both positive, then θ is between 0 and $\frac{\pi}{2}$; if x is negative and y is positive, then θ is between $\frac{\pi}{2}$ and π; if x and y are both negative, then θ is between π and $\frac{3\pi}{2}$; and if x is positive and y is negative, then θ is between $\frac{3\pi}{2}$ and 2π. To convert from polar coordinates back into Cartesian coordinates, one simply takes

$$x = r\cos\theta$$

$$y = r\sin\theta$$

Note that because of the periodicity of the sine and cosine function, there are many values of θ that will correspond to the same point. Specifically, $\theta' = \theta + 2k\pi$ yields the same point for any integer value of k.

Test Yourself

In problems 1–10, prove the identity.

1. $\cos(2x) = \dfrac{1 - \tan^2(x)}{1 + \tan^2(x)}$

2. $\tan(2x) = \dfrac{2\tan(x)}{1 - \tan^2(x)}$

3. $\cot^2(x) + \cos^2(x) + \sin^2(x) = \csc^2(x)$

4. $\tan^2(x) - \sin^2(x) = \tan^2(x)\sin^2(x)$

5. $\dfrac{\sin(t)}{\sin(t) + \cos(t)} = \dfrac{\tan(t)}{1 + \tan(t)}$

6. $\dfrac{\cot(x) - \tan(x)}{\sin(x)\cos(x)} = \csc^2(x) - \sec^2(x)$

7. $\dfrac{1 - \sin^2(x)}{1 + \cot^2(x)} = \sin^2(x)\cos^2(x)$

8. $\dfrac{\tan^2(x)}{1 + \tan^2(x)} = \sin^2(x)$

9. $\dfrac{\tan(x)}{1 + \sec(x)} + \dfrac{1 + \sec(x)}{\tan(x)} = 2\csc(x)$

10. $\sin(3x) = 3\sin(x) - 4\sin^3(x)$

In problems 11–20, either prove the expression is an identity or verify that it is not.

11. $\sin(x) + \cos(x) = \dfrac{\cos(2x)}{[\cos(x) - \sin(x)]}$

12. $\sin(x) + \cos(x) = \sqrt{2}\sin(45° + x)$

13. $\sec(x) - \csc(x) = \tan(2x)$

14. $\sin(4x) = 4\sin(x)\cos(x) - 8\sin^3(x)\cos(x)$

15. $\sin^2(x) = \dfrac{[1 - \cos(2x)]}{2}$

16. $[\sin^3(x) - \sin(x)]\sec^2(x) = -\sin(x)$

17. $\tan^2(x)\cos(x) = \sec(x) - \cos(x)$

18. $\cot^2(x)\sin(x) = \csc(x)$

19. $\cot(y) + \cot(x) = \dfrac{\sin(x + y)}{[\sin(x)\sin(y)]}$

20. $\tan(x)\cos(x)\csc(x) = \sin^2(x) + \cos^2(x)$

Solve the trigonometric equations given in problems 21–28 using degree measure. Give all values for x such that $0 \le x < 360°$ for which the equation is satisfied.

21. $8\cos^2(x) - 3 = 1$

22. $\cos^2(x) - 3\sin(x) = 3$

23. $6\sin^2(x) - 7\sin(x) + 2 = 0$

24. $\sec^2(x) = 25$

25. $1 - \csc^2(x) = -8$

26. $6\sin^2(x) = 7 - 5\cos(x)$

27. $\tan^2(x) = 36$

28. $2\tan^2(x) = 3\tan(x) - 1$

Solve the trigonometric equations given in problems 29–36 in terms of radian measure. Give all solutions for x such that $0 \le x < 2\pi$.

29. $\sec(x) = 2\csc(x)$

30. $\sec(x)\sin(x) = 2\sin(x)$

31. $\tan^2(x) = \tan(x)$

32. $\sin^2(x) = \sin(x)$

33. $\sin(x) + \cos(x) = 0$

34. $1 + 2\cos^2(x) = \sin(x)$

35. $2\sin(x)\cos(x) = \tan(x)$

36. $\sin^2(x) + \sin(x) - 1 = 0$

Convert the complex numbers in problems 37–41 into *cis* notation.

37. $z = 1 + i$

38. $z = 3 - 4i$

39. $z = -1 - 2i$

40. $z = \dfrac{1}{\sqrt{2}} + \dfrac{i}{\sqrt{2}}$

41. $z = -5 + 12i$

Convert the expressions in problems 42–46 into $z = x + iy$ notation.

42. $\sqrt{2}\,\text{cis}(45°)$

43. $5\text{cis}(53.13°)$

44. $7\text{cis}(225°)$

45. $12\text{cis}(330°)$

46. $2\text{cis}(150°)$

Find both square roots of the complex numbers in problems 47–49.

47. i

48. $2\text{cis}(120°)$

49. $\dfrac{\sqrt{3}}{2} + \dfrac{i}{2}$

Find all three cube roots of the complex numbers in problems 50–53.

50. i

51. $8\text{cis}(120°)$

52. $1 + i$

53. Use De Moivre's Theorem to prove the identities
$\sin(5x) = 5\sin(x) - 20\sin^3(x) + 16\sin^5(x)$ and
$\cos(5x) = 5\cos(x) - 20\cos^3(x) + 16\cos^5(x)$.

 # Check Yourself

1. Working on the right side, we have $\dfrac{1 - \tan^2(x)}{1 + \tan^2(x)} = \dfrac{1 - \dfrac{\sin^2 x}{\cos^2 x}}{1 + \dfrac{\sin^2 x}{\cos^2 x}} = \dfrac{\cos^2 x\left(\dfrac{1 - \sin^2 x}{\cos^2 x}\right)}{\cos^2 x\left(1 + \dfrac{\sin^2 x}{\cos^2 x}\right)} =$

$\dfrac{[\cos^2(x) - \sin^2(x)]}{[\cos^2(x) + \sin^2(x)]} = \dfrac{\cos(2x)}{1} = \cos(2x)$. (**Trigonometric identities**)

2. Working on the right side, we have $\dfrac{[2\tan(x)]}{[1 - \tan^2(x)]} = \dfrac{\left[\dfrac{2\sin(x)}{\cos(x)}\right]}{1 - \dfrac{\sin^2 x}{\cos^2 x}} = \dfrac{\cos^2(x)\left[\dfrac{2\sin(x)}{\cos(x)}\right]}{\cos^2(x)\left[1 - \dfrac{\sin^2(x)}{\cos^2(x)}\right]} =$

$\dfrac{2\sin(x)\cos(x)}{[\cos^2(x) - \sin^2(x)]} = \dfrac{\sin(2x)}{\cos(2x)} = \tan(2x)$. (**Trigonometric identities**)

3. Working on the left side, we have $\cot^2(x) + \cos^2(x) + \sin^2(x) = \dfrac{\cos^2(x)}{\sin^2(x)} + 1 =$

   $\dfrac{[\cos^2(x) + \sin^2(x)]}{\sin^2(x)} = \dfrac{1}{\sin^2(x)} = \csc^2(x).$ **(Trigonometric identities)**

4. Working on the right side, we have $\tan^2(x)\sin^2(x) = \tan^2(x)[1 - \cos^2(x)] = \tan^2(x) - \tan^2(x)\cos^2(x) =$

   $\tan^2(x) - \dfrac{[\sin^2(x)\cos^2(x)]}{\cos^2(x)} = \tan^2(x) - \sin^2(x).$ **(Trigonometric identities)**

5. Working on the right side we have $\dfrac{\tan(t)}{1 + \tan(t)} = \left(\dfrac{\frac{\sin(t)}{\cos(t)}}{1 + \frac{\sin(t)}{\cos(t)}}\right) = \left(\dfrac{\frac{\sin(t)}{\cos(t)}}{\frac{\cos(t)}{\cos(t)} + \frac{\sin(t)}{\cos(t)}}\right) = \left(\dfrac{\frac{\sin(t)}{\cos(t)}}{\frac{\cos(t) + \sin(t)}{\cos(t)}}\right) =$

   $\dfrac{\sin(t)}{\cos(t) + \sin(t)}.$ **(Trigonometric identities)**

6. Working on the left side, we have $\dfrac{\cot(x) - \tan(x)}{\sin(x)\cos(x)} = \dfrac{\sin(x)[\cot(x) - \tan(x)]}{\sin(x)[\sin(x)\cos(x)]} = \dfrac{\cos(x)}{\sin^2(x)\cos(x)} -$

   $\dfrac{\sin(x)}{\sin(x)\cos^2(x)} = \dfrac{1}{\sin^2(x)} - \dfrac{1}{\cos^2(x)} = \csc^2(x) - \sec^2(x).$ **(Trigonometric identities)**

7. Working on the left side we have $\dfrac{1 - \sin^2(x)}{1 + \cot^2(x)} = \dfrac{\cos^2(x)}{1 + \frac{\cos^2(x)}{\sin^2(x)}} = \dfrac{\cos^2(x)}{\frac{\sin^2(x) + \cos^2(x)}{\sin^2(x)}} = \dfrac{\cos^2(x)\sin^2(x)}{\sin^2(x) + \cos^2(x)} =$

   $\cos^2(x)\sin^2(x).$ **(Trigonometric identities)**

8. Working on the left side, we have $\dfrac{\tan^2(x)}{1 + \tan^2(x)} = \dfrac{\frac{\sin^2(x)}{\cos^2(x)}}{1 + \frac{\sin^2 x}{\cos^2 x}} = \dfrac{\frac{\sin^2(x)}{\cos^2(x)}}{\frac{\cos^2(x) + \sin^2(x)}{\cos^2(x)}} = \dfrac{\sin^2(x)}{\cos^2(x) + \sin^2(x)} = \sin^2(x).$

   (Trigonometric identities)

9. Working on the left side, we have

   $\dfrac{\tan(x)}{1 + \sec(x)} + \dfrac{1 + \sec(x)}{\tan(x)} = \dfrac{\frac{\sin(x)}{\cos(x)}}{1 + \frac{1}{\cos(x)}} + \dfrac{1 + \frac{1}{\cos(x)}}{\frac{\sin(x)}{\cos(x)}} = \dfrac{\frac{\sin(x)}{\cos(x)}}{\frac{1 + \cos(x)}{\cos(x)}} + \dfrac{\frac{\cos(x) + 1}{\cos(x)}}{\frac{\sin(x)}{\cos(x)}} = \dfrac{\sin(x)}{1 + \cos(x)} + \dfrac{\cos(x) + 1}{\sin(x)} =$

   $\dfrac{\sin^2(x) + 1 + 2\cos(x) + \cos^2(x)}{\sin(x)(1 + \cos(x))} = \dfrac{2 + 2\cos(x)}{\sin(x)(1 + \cos(x))} = \dfrac{2[1 + \cos(x)]}{\sin(x)(1 + \cos(x))} = \dfrac{2}{\sin(x)} = 2\csc(x).$
   (Trigonometric identities)

10. This problem is an exception to the rule that it is usually best to work on the most complex side. Working on the left side, we have

    $\sin(3x) = \sin(2x + x) + \sin(2x)\cos(x) + \sin(x)\cos(2x) = 2\sin(x)\cos^2(x) + \sin(x)[\cos^2(x) - \sin^2(x)] =$
    $3\sin(x)\cos^2(x) - \sin^3(x) = 3\sin(x)[1 - \sin^2(x)] - \sin^3(x) = 3\sin(x) - 4\sin^3(x).$ **(Trigonometric identities)**

11. Working on the right side, we have $\dfrac{\cos(2x)}{\cos(x) - \sin(x)} = \dfrac{\cos^2(x) - \sin^2(x)}{\cos(x) - \sin(x)} =$

    $\dfrac{[\cos(x) + \sin(x)][\cos(x) - \sin(x)]}{\cos(x) - \sin(x)} = \cos(x) + \sin(x)$. (**Trigonometric identities**)

12. Apply the addition formula to the right side and the result follows: $\sqrt{2}\sin(45° + x) =$

    $\sqrt{2}[\sin(45°)\cos(x) + \sin(x)\cos(45°)] = \sqrt{2}\left[\dfrac{1}{\sqrt{2}}\cos(x) + \dfrac{1}{\sqrt{2}}\sin(x)\right] = \cos(x) + \sin(x).$

    (**Trigonometric identities**)

13. Replace x by $30°$ and the left side becomes negative while the right side becomes positive, so the expression is not an identity. (**Trigonometric identities**)

14. Working on the left side, $\sin(4x) = 2\sin(2x)\cos(2x) = 4\sin(x)\cos(x)[\cos^2(x) - \sin^2(x)] =$
    $4\sin(x)\cos(x)[1 - 2\sin^2(x)] = 4\sin(x)\cos(x) - 8\sin^3(x)\cos(x).$ (**Trigonometric identities**)

15. Working on the right side, $\dfrac{[1 - \cos(2x)]}{2} = \dfrac{[1 - \cos^2(x) + \sin^2(x)]}{2} = \dfrac{2\sin^2(x)}{2} = \sin^2(x).$ (**Trigonometric identities**)

16. Working on the left side, we have $[\sin^3(x) - \sin(x)]\sec^2(x) = \dfrac{\sin(x)(\sin^2(x) - 1)}{\cos^2(x)} = \dfrac{\sin(x)(-\cos^2(x))}{\cos^2(x)} = -\sin(x).$

    (**Trigonometric identities**)

17. Working on the right side, we have $\sec(x) - \cos(x) = \dfrac{1}{\cos(x)} - \cos(x) = \dfrac{1 - \cos^2(x)}{\cos(x)} = \dfrac{\sin^2(x)}{\cos(x)} =$

    $\cos(x)\dfrac{\sin^2(x)}{\cos^2(x)} = \cos(x)\tan^2(x).$ (**Trigonometric identities**)

18. Substitute $45°$ for x and you get $\dfrac{\sqrt{2}}{2} = \sqrt{2}$, which is not true, so the equation is not an identity.

    (**Trigonometric identities**)

19. Working on the right side, use the addition formula for sines, and then apply the identity $\dfrac{\cos(u)}{\sin(u)} = \cot(u).$

    $\dfrac{\sin(x + y)}{[\sin(x)\sin(y)]} = \dfrac{\sin(x)\cos(y) + \sin(y)\cos(x)}{\sin(x)\sin(y)} = \dfrac{\sin(x)\cos(y)}{\sin(x)\sin(y)} + \dfrac{\sin(y)\cos(x)}{\sin(x)\sin(y)} = \dfrac{\cos(y)}{\sin(y)} + \dfrac{\cos(x)}{\sin(x)} = \cot(y) + \cot(x)$
    (**Trigonometric identities**)

20. Both sides of this expression are equal to 1, so it is an identity.
    $\tan(x)\cos(x)\csc(x) = \dfrac{\sin(x)}{\cos(x)} \cdot \cos(x) \cdot \dfrac{1}{\sin(x)} = 1 = \sin^2(x) + \cos^2(x)$ (**Trigonometric identities**)

21. $8\cos^2(x) - 3 = 1$
    $8\cos^2(x) = 4$
    $\cos^2(x) = \dfrac{1}{2}$ This is equivalent to $\cos(x) = \dfrac{1}{\sqrt{2}}$ or $\dfrac{-1}{\sqrt{2}}$. So $x = 45°, 135°, 225°,$ or $315°.$
    (**Trigonometric equations**)

22. Replace $\cos^2(x)$ by $1 - \sin^2(x)$, and solve the resulting quadratic.
$1 - \sin^2(x) - 3\sin(x) = 3$
$\sin^2(x) + 3\sin(x) + 2 = 0$
$[\sin(x) + 2][\sin(x) + 1] = 0$
You get $\sin(x) = -2$ or $\sin(x) = -1$. The former is true for no values of x, so the only solution is $x = 270°$. (**Trigonometric equations**)

23. Begin by factoring the quadratic equation: $[3\sin(x) - 2][2\sin(x) - 1] = 0$, so $\sin(x) = \frac{2}{3}$ or $\sin(x) = \frac{1}{2}$. That gives us $x = 41.8°$, $138.2°$, $30°$ and $150°$. (**Trigonometric equations**)

24. Take the square root of both sides: $\sec(x) = 5$ or $\sec(x) = -5$. That gives us $x = 78.5°$, $281.5°$, $101.5°$, and $258.5°$. (**Trigonometric equations**)

25. $1 - \csc^2(x) = -8$
$-\csc^2(x) = -9$
$\csc^2(x) = 9$
$\csc(x) = \pm 3$, so
$\sin(x) = \frac{1}{3}$ or $-\frac{1}{3}$. This gives us $x = 19.5$, 340.5, $160.5°$, and $199.5°$. (**Trigonometric equations**)

26. Use the Pythagorean identity to create an equation in $\cos(x)$, and you get $[2\cos(x) - 1][3\cos(x) - 1] = 0$. Therefore $\cos(x) = \frac{1}{2}$ or $\frac{1}{3}$, which gives us $x = 60°$, $300°$, $70.5°$, and $289.5°$. (**Trigonometric equations**)

27. Take the square root of both sides: $\tan(x) = +6$ or $\tan(x) = -6$. That gives us $x = 80.5°$, $260.5°$, $279.5°$, and $99.5°$. (**Trigonometric equations**)

28. $[2\tan)(x) - 1][\tan(x) - 1] = 0$, so either $\tan(x) = \frac{1}{2}$ or $\tan(x) = 1$. That gives us $x = 26.6°$, $206.6°$, $45°$, and $225°$. (**Trigonometric equations**)

29. $\sec(x) = 2\csc(x)$
$\dfrac{\sec(x)}{\csc(x)} = 2$
$\tan(x) = 2$, so $x = 1.107$ or 4.249. (**Trigonometric equations**)

30. $\sec(x)\sin(x) = 2\sin(x)$
$2\sin(x) - \sec(x)\sin(x) = 0$
$\sin(x)(2 - \sec(x)) = 0$, so
We have $\sec(x) = 2$ or $\sin(x) = 0$, so $x = 0$, 1.047, 3.142, or 5.236. (**Trigonometric equations**)

31. $\tan^2(x) = \tan(x)$
$\tan^2(x) - \tan(x) = 0$
$\tan(x)(\tan(x) - 1) = 0$, so
$\tan(x) = 1$ or $\tan(x) = 0$, so $x = 0$, π, $\dfrac{\pi}{4}$, or $\dfrac{5\pi}{4}$. (**Trigonometric equations**)

32. $\sin(x) = 1$ or $\sin(x) = 0$, so $x = 0$, $\dfrac{\pi}{2}$, or π. (**Trigonometric equations**)

33. $\sin(x) + \cos(x) = 0$

$\sin(x) = -\cos(x)$

$\dfrac{\sin(x)}{\cos(x)} = \dfrac{-\cos(x)}{\cos(x)}$

$\tan(x) = -1$, so $x = \dfrac{7\pi}{4}, \dfrac{3\pi}{4}$ (**Trigonometric equations**)

34. Use the Pythagorean identity to put everything in terms of the sine function, and you get

$1 + 2\cos^2(x) = \sin(x)$

$1 + 2[1 - \sin^2(x)] = \sin(x)$

$1 + 2 - 2\sin^2(x) = \sin(x)$

$2\sin^2(x) + \sin(x) - 3 = 0$

$[2\sin(x) + 3][\sin(x) - 1] = 0$. There are no values of x such that $\sin(x) = -\dfrac{3}{2}$, so the only solution consists

of values of x such that $\sin(x) = 1$. We have $x = \dfrac{\pi}{2}$ as the only solution. (**Trigonometric equations**)

35. $2\sin(x)\cos(x) = \tan(x)$

$2\sin(x)\cos(x) = \dfrac{\sin(x)}{\cos(x)}$

$2\sin(x)\cos^2(x) = \sin(x)$

$2\sin(x)\cos^2(x) - \sin(x) = 0$

$\sin(x)(2\cos^2(x) - 1) = 0$

So we either have $\sin(x) = 0$ or $\cos(x) = \dfrac{1}{\sqrt{2}}$ or $\cos(x) = \dfrac{-1}{\sqrt{2}}$. Therefore we must have

$x = 0, \pi, \dfrac{\pi}{4}, \dfrac{3\pi}{4}, \dfrac{5\pi}{4}, \dfrac{7\pi}{4}$. (**Trigonometric equations**)

36. Using the quadratic formula, we find that $\sin(x) = \dfrac{-1 - \sqrt{5}}{2}$ or $\sin(x) = \dfrac{-1 + \sqrt{5}}{2}$. The first quantity is less

than -1, so it gives no solutions. From the second quantity, we get $x = 0.6662$ and 2.4754. (**Trigonometric equations**)

37. In this problem, $x = y = 1$. Thus, $r = \sqrt{x^2 + y^2} = \sqrt{1 + 1} = \sqrt{2}$. $\beta = \arctan\left(\dfrac{y}{x}\right) = \arctan(1) = 45°$. Thus,

$1 + i = \sqrt{2}\,\text{cis}(45°)$. (**Applications to complex numbers**)

38. $5\text{cis}(306.87°)$ (**Applications to complex numbers**)

39. $\sqrt{5}\,\text{cis}(243.43°)$ (**Applications to complex numbers**)

40. $\text{cis}(45°)$ (**Applications to complex numbers**)

41. $13\text{cis}(112.62°)$ (**Applications to complex numbers**)

42. $\sqrt{2}\,\text{cis}(45°) = \sqrt{2}\,[\sin(45°) + i\cos(45°)] = \sqrt{2}\left(\dfrac{1}{\sqrt{2}} + i \cdot \dfrac{1}{\sqrt{2}}\right) = 1 + i$ (**Applications to complex numbers**)

43. $3 + 4i$ (**Applications to complex numbers**)

44. $\dfrac{-7}{\sqrt{2}} + \dfrac{-7i}{\sqrt{2}}$ **(Applications to complex numbers)**

45. $6\sqrt{3} - 6i$ **(Applications to complex numbers)**

46. $-\sqrt{3} + i$ **(Applications to complex numbers)**

47. By DeMoivre's Formula, $i = 1[\cos(90°) + i\sin(90°)]$. Then, $i^{\frac{1}{2}} = 1^{\frac{1}{2}}\left[\cos\left(\dfrac{90°}{2}\right) + i\sin\left(\dfrac{90°}{2}\right)\right] = $
$1(\cos 45° + i\sin 45°) = \dfrac{1}{\sqrt{2}} + \dfrac{i}{\sqrt{2}}$. Also, $i^{\frac{1}{2}} = 1^{\frac{1}{2}}\left[\cos\left(\dfrac{450°}{2}\right) + i\sin\left(\dfrac{450°}{2}\right)\right] = 1(\cos 225° + i\sin 225°) = -\dfrac{1}{\sqrt{2}} - \dfrac{i}{\sqrt{2}}$.
Thus, the solutions are $\dfrac{1}{\sqrt{2}} + \dfrac{i}{\sqrt{2}}$ and $\dfrac{-1}{\sqrt{2}} - \dfrac{i}{\sqrt{2}}$. **(Applications to complex numbers)**

48. $\sqrt{2}\,\text{cis}(60°)$ and $\sqrt{2}\,\text{cis}(240°)$ **(Applications to complex numbers)**

49. $\text{cis}(15°)$ and $\text{cis}(195°)$ **(Applications to complex numbers)**

50. Begin by expressing i as $1[\cos(90°) + i\sin(90°)]$. Then the three cube roots are
$1^{\frac{1}{3}}[\cos(30°) + i\sin(30°)] = \text{cis}(30°)$ and

$i^{\frac{1}{3}} = 1^{\frac{1}{3}}\left[\cos\left(\dfrac{450°}{3}\right) + i\sin\left(\dfrac{450°}{3}\right)\right] = 1\text{cis}(150°)$ and

$i^{\frac{1}{3}} = 1^{\frac{1}{3}}\left[\cos\left(\dfrac{810°}{3}\right) + i\sin\left(\dfrac{810°}{3}\right)\right] = 1\text{cis}(270°)$. Thus, the roots are $\text{cis}(30°)$, $\text{cis}(150°)$, and $\text{cis}(270°)$.

(Applications to complex numbers)

51. $2\text{cis}(40°)$, $2\text{cis}(160°)$, and $2\text{cis}(280°)$. **(Applications to complex numbers)**

52. $(2^{(\frac{1}{6})})\text{cis}(15°)$, $(2^{(\frac{1}{6})})\text{cis}(135°)$, and $(2^{(\frac{1}{6})})\text{cis}(255°)$. **(Applications to complex numbers)**

53. Consider the fifth power of $\cos(x) + i\sin(x)$. By De Moivre's Formula, this will be $\cos(5x) + i\sin(5x)$. By multiplying out in the usual way, we obtain $[\cos^5(x) - 10\cos^3(x)\sin^2(x) + 5\cos(x)\sin^4(x)] + i[\sin^5(x) - 10\sin^3(x)\cos^2(x) + 5\sin(x)\cos^4(x)]$. Use the Pythagorean identities to put each of these into terms of $\cos(x)$ and $\sin(x)$ respectively and the desired result follows. **(Applications to complex numbers)**

Grade Yourself

Circle the numbers of the questions you missed, then fill in the total incorrect for each topic. If you answered more than three questions incorrectly, you need to focus on that topic. (If a topic has fewer than three questions and you had at least one wrong, we suggest you study that topic also. Read your textbook or a review book, or ask your teacher for help.)

Subject: Trigonometry: Identities, Equations, and Applications

Topic	Question Numbers	Number Incorrect
Trigonometric identities	1, 2, 3, 4, 5, 6, 7, 8, 9, 10, 11, 12, 13, 14, 15, 16, 17, 18, 19, 20	
Trignonometric equations	21, 22, 23, 24, 25, 26, 27, 28, 29, 30, 31, 32, 33, 34, 35, 36	
Applications to complex numbers	37, 38, 39, 40, 41, 42, 43, 44, 45, 46, 47, 48, 49, 50, 51, 52, 53	

Systems of Equations

Brief Yourself

A *system of equations* is a set of two or more equations for which we wish to find a set of values that makes all of the equations true at the same time. Systems of equations are also called *simultaneous equations*, because we are interested in finding values for the variables that solve all of the equations at the same time. Being able to solve a system of simultaneous equations is very important, because there are many types of problems in mathematics in which we have a number of unknown quantities and a given set of relationships among these unknowns. In this chapter, we review a number of ways by which we can solve systems of equations.

An equation of the form $ax + by = c$, where a, b, and c are real numbers and a and b are not both 0, is called a *linear equation in two variables*. A set of two such equations is called a *system of two linear equations in two variables*. When finding the common solution of such a system, there are three possible outcomes: the system has a unique solution (*independent system*), the system has no solution (*inconsistent system*), or the system has more than one solution (*dependent system*).

The graph of a linear equation in two variables is a straight line. Thus, one way to solve a system of two equations in two variables (a 2×2 system) is to graph both equations on a coordinate axis. If the two lines intersect in one point, the coordinates of that point represent the common solution. If the two lines are parallel, we have an inconsistent system and no solutions; if the two lines coincide, we have a dependent system and an infinite amount of solutions.

This graphical method of solution has a major drawback in that in most cases it is not possible to accurately read off the point of intersection. Algebraic methods are more reliable. There are two common algebraic methods: the method of *elimination* (sometimes called Gaussian elimination) and the method of *substitution*. To use the method of elimination, the equations are transformed in such a way that when they are subsequently added together, one of the unknowns cancels out and we can solve for the other. To use the method of substitution, solve one of the equations for one of the unknowns and plug the value obtained into the other equation. These two solution techniques can also be used to solve systems of linear equations in more than two unknowns, although the algebra becomes more and more cumbersome as the number of variables increases.

There are several other techniques that can be used to solve systems of equations. One involves using what are known as *determinants*. For any real numbers a_1, b_1, a_2, and b_2, the determinant $D = \begin{vmatrix} a_1 & b_1 \\ a_2 & b_2 \end{vmatrix}$ is defined to have the value $a_1b_2 - a_2b_1$. Now, consider the following system of equations:

$$a_1 x + b_1 y = c_1$$

$$a_2 x + b_2 y = c_2$$

It can be shown that if we define the following determinants $D = \begin{vmatrix} a_1 & b_1 \\ a_2 & b_2 \end{vmatrix}$, $D_x = \begin{vmatrix} c_1 & b_1 \\ c_2 & b_2 \end{vmatrix}$, and

$D_y = \begin{vmatrix} a_1 & c_1 \\ a_2 & c_2 \end{vmatrix}$ and if $D \neq 0$, the common solution of the equations is given by

$$x = \frac{D_x}{D} \text{ and } y = \frac{D_y}{D}$$

The solution of a system of linear equations in determinant form is known as *Cramer's Rule*. The definition of determinants can be generalized to 3×3 determinants and larger, and Cramer's Rule can be generalized to provide formulas for the solutions of such systems. In general, however, Cramer's Rule is very time consuming for systems of high order since it involves evaluating many determinants of high order. Thus, we consider one more technique for solving systems of equations.

The method of elimination that was discussed above can be made much more efficient by the use of matrices. A *matrix* is simply an array of numbers; matrices can be used to hold all of the numbers in a system of equations. For example, consider the system.

$$2x - 3y + z = 1$$

$$x - 4y - z = 1$$

$$x + 2y + 3z = 4$$

The left-hand side of this equation can be represented by the matrix of coefficients:

$$M = \begin{pmatrix} 2 & -3 & 1 \\ 1 & -4 & -1 \\ 1 & 2 & 3 \end{pmatrix}$$

In order to include the right-hand side of the system, we can expand the matrix to form

$$M^* = \begin{pmatrix} 2 & -3 & 1 & | & 1 \\ 1 & -4 & -1 & | & 1 \\ 1 & 2 & 3 & | & 4 \end{pmatrix}$$

which is called an *augmented matrix*. By performing elementary row operations on the augmented matrix, it can be changed to echelon form, from which the solution can easily be found.

Test Yourself

1. Verify that the ordered pair (3, 2) belongs to the solution set of the following system of equations:

   $2x + 3y = 12$

   $6x - 2y = 14$

2. Verify that the ordered triple (1, 2, 3) belongs to the solution set of the following system of equations:

   $x + y + z = 6$

   $2x - y + z = 3$

   $3x + y - z = 2$

3. Describe the three possible outcomes when solving a system of linear equations.

4. Define an independent system of equations, an inconsistent system of equations, and a dependent system of equations.

5. For the case of a system of two linear equations in two unknowns, describe the graph of an independent system, an inconsistent system, and a dependent system.

6. Solve the system of equations given below by graphing both equations:

   $2x - 3y = 12$

   $4x + 5y = 2$

7. Solve the system of equations given below by graphing both equations:

   $4x + 3y = 15$

   $-x + 4y = 20$

8. Solve the system of equations given below by graphing both equations.

   $3x - 2y = 10$

   $-6x + 4y = 8$

9. Solve the system of equations given below by graphing both equations:

   $2x - 8y = 6$

   $4x - 16y = 12$

10. When using the method of elimination to solve a 2×2 system of equations, what will the result be if the system happens to be inconsistent?

11. When using the method of elimination to solve a 2×2 system of equations, what will the result be if the system happens to be dependent?

12. Solve the system of equations given below by using the method of elimination:

    $2x + y = 5$

    $x - y = 1$

13. Solve the system of equations given below by using the method of elimination:

    $2x + 3y = 7$

    $x + y = 2$

14. Solve the system of equations given below by using the method of elimination:

    $3x + 2y = 8$

    $2x - 3y = 14$

15. Solve the system of equations given below by using the method of elimination:

    $3x + y = p$

    $x + y = q$

16. Solve the system of equations given below by using the method of elimination:

    $3x - 2y = 10$

    $-6x + 4y = 8$

17. Solve the system of equations given below by using the method of elimination:

    $2x - 8y = 6$

    $4x - 16y = 12$

18. Under what circumstances might it be easier to solve a system of equations by using the method of substitution instead of the method of elimination?

In problems 19–22, find the solution set of each system by the method of substitution.

19. $y = x - 2$
    $2y = x - 3$

20. $y = 3x + 2$
    $x + y = 6$

21. $2x - y = 3$
    $3x + y = 22$

22. $3x + y = 4$
    $7x - y = 6$

In problems 23 and 24, use the method of elimination to find the solution set of the given system.

23. $x - 2y + 3z = 10$
    $x + 2y - z = 6$
    $-2x + y + 2z = -2$

24. $x - y + z = 4$
    $3x + y + 2z = 20$
    $4x - y + 5z = 29$

In problems 25–28, solve the word problems for the unknowns indicated.

25. The sum of two numbers is 27 and their difference is 11. What are the numbers?

26. A 24-foot board is cut into two pieces, one of which is 2 feet longer than the other. What is the length of the shorter piece?

27. In a recent election, the winning candidate received 122 more votes than her opponent. If there were a total of 10,764 votes cast, how many votes did each candidate receive?

28. Brian has a pile of coins containing 34 dimes and quarters. How many coins of each kind does he have if the total value of the coins is $5.50?

29. What is the determinant of a matrix that contains a row of 0's? a column of 0's?

30. What is the determinant of a matrix that has two rows that are the same?

31. Given any 2×2 matrix, explain how to compute the value of its determinant.

In problems 32–35, find the value of the given determinants.

32. $\begin{vmatrix} 10 & 7 \\ 2 & -7 \end{vmatrix}$

33. $\begin{vmatrix} -5 & 6 \\ -3 & 4 \end{vmatrix}$

34. $\begin{vmatrix} 6 & 8 \\ 3 & 4 \end{vmatrix}$

35. $\begin{vmatrix} 13 & 25 \\ 0 & 0 \end{vmatrix}$

36. State Cramer's Rule for the solution of two linear equations in two variables. Prove that Cramer's Rule gives the correct solution in the case of a 2×2 system of equations.

37. Describe the outcome of using Cramer's Rule to solve a system of inconsistent equations.

38. Describe the outcome of using Cramer's Rule to solve a system of dependent equations.

In problems 39–42, use Cramer's Rule to solve the given systems of equations.

39. $2x - 3y = 12$
    $4x + 5y = 2$

40. $4x + 3y = 15$
    $-x + 4y = 20$

41. $3x - 2y = 10$
    $-6x + 4y = 8$

42. $2x - 8y = 6$
    $4x - 16y = 12$

In problems 43–46, evaluate the determinants.

43. $\begin{vmatrix} 0 & 0 & 3 \\ 0 & 5 & 0 \\ -2 & 4 & 0 \end{vmatrix}$

44. $\begin{vmatrix} 5 & 6 & 0 \\ -4 & 0 & 2 \\ 1 & -1 & 1 \end{vmatrix}$

45. $\begin{vmatrix} 2 & 9 & 7 \\ 0 & 0 & 0 \\ 5 & 2 & 7 \end{vmatrix}$

46. $\begin{vmatrix} 4 & 2 & 5 \\ -1 & 4 & -2 \\ 4 & 2 & 5 \end{vmatrix}$

In problems 47–48, use Cramer's Rule to solve the given systems of equations.

47. $x - 2y + 3z = 10$
$x + 2y - z = 6$
$-2x + y + 2z = -2$

48. $x - y + z = 4$
$3x + y + 2z = 20$
$4x - y + 5z = 29$

49. What is a matrix? How can a matrix be used to represent the coefficients of the variables in a system of equations?

50. What is an augmented matrix? How can an augmented matrix be used to represent an entire system of equations?

51. Express the system of equations given in problem 12 in an augmented matrix.

52. Express the system of equations given in problem 47 in an augmented matrix.

53. What are the three elementary matrix row operations?

54. What does it mean for two matrices to be row equivalent?

55. What does it mean for a matrix to be in echelon form?

56. Use matrices to solve the system of equations given in problem 12.

57. Use matrices to solve the system of equations given in problem 16.

58. Use matrices to solve the system of equations given in problem 22.

59. Use matrices to solve the system of equations given in problem 23.

Find the solutions to word problems 60–64, using the technique of your choice to solve the equations you obtain.

60. Find two numbers whose difference is 67 and whose product is 3,300.

61. A sum of $2000 is invested, part at 7% and the rest at 8%. How much was invested at each rate if the yearly interest from the two investments is $151?

62. One number is eight more than four times the other, and their sum is −2. What are the numbers?

63. John is three years older than Dennis. Four years ago he was twice as old as Dennis. How old is each now?

64. The Flansbergh Company produces two different types of computer printers. It costs the company $100 to produce a Type A printer and $150 to produce a Type B printer. During the month of January, the total number of printers of both types produced was 55, and the total cost of production was $7000. How many printers of each type were produced?

Check Yourself

1. We simply need to plug (3, 2) into both equations and check if this point solves the equations.
    $2x + 3y = 2(3) + 3(2) = 6 + 6 = 12$
    $6x - 2y = 6(3) - 2(2) = 18 - 4 = 14$ Thus, (3, 2) belongs to the solution set of the system of equations.
    (Checking the solution of a system of equations)

2. Check to see if (1, 2, 3) solves all three equations:
    $x + y + z = 1 + 2 + 3 = 6$
    $2x - y + z = 2(1) - 2 + 3 = 3$
    $3x + y - z = 3(1) + 2 - 3 = 2$ Thus (1, 2, 3) belongs to the solution set of the system of equations.
    (Checking the solution of a system of equations)

3. The most common outcome is that the system has a single unique solution. It is also possible, however, for the system to have no solution at all or an infinite number of solutions. **(Solving systems of equations—theory)**

4. If a system of equations has a single unique solution, it is called an independent system. If it has no solutions at all, it is called an inconsistent system, and if it has an infinite number of solutions, it is called a dependent system. **(Solving systems of equations—theory)**

5. The graph of an independent system consists of two straight lines that intersect in a single point. The graph of an inconsistent system consists of two parallel lines that, of course, never intersect. The graph of a dependent system consists of two coincident lines. **(Solving systems of equations—theory)**

6. Since both equations are linear, the graphs of both will be straight lines. Thus, we need to find only two points that solve each equation in order to graph the lines. The quickest way to find two solutions for a given equation is to use the intercept method. First, let $x = 0$, and solve for y. Then, let $y = 0$ and solve for x. In the case of $2x - 3y = 12$, when $x = 0$, we can see y must be –4. Similarly, when $y = 0$, we have $x = 6$. Thus, our two solutions are (0, –4) and (6, 0). Similarly, for the second equation $4x + 5y = 2$, the solutions are $(0, \frac{2}{5})$ and $(\frac{1}{2}, 0)$. Now, graph the two lines on the same axis (Figure 7-1):

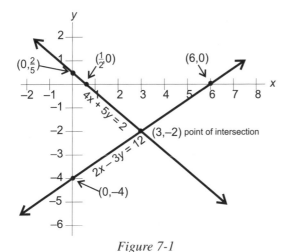

Figure 7-1

From the graph, it is clear that the common solution is (3, –2). **(Solving systems of equations by graphing)**

7. To graph $4x + 3y = 15$, let's graph the points (3, 1) and (6, –3). For $-x + 4y = 20$, let's graph (4, 6) and (–4, 4). Then, it can be seen from the graph in Figure 7-2 that the common solution is (0, 5).

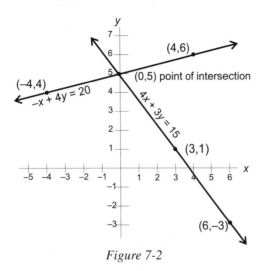

Figure 7-2

(Solving systems of equations by graphing)

8. To graph $3x – 2y = 10$, plot (0, –5) and (2, –2). To graph $-6x + 4y = 8$, plot (0, 2) and ($\frac{-4}{3}$, 0) (Figure 7-3):

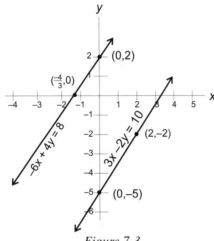

Figure 7-3

The lines are parallel. There is no common solution. The system of equations is inconsistent. **(Solving systems of equations by graphing)**

9. For $2x - 8y = 6$, plot (3, 0) and (–1,–1). For $4x - 16y = 12$, plot $(0, \frac{-3}{4})$ and $(2, \frac{-1}{4})$ (Figure 7-4).

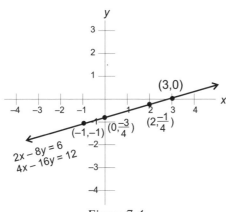

Figure 7-4

The two lines are the same. There are an infinite number of solutions. The system is dependent. **(Solving systems of equations by graphing)**

10. If the system of equations is inconsistent, the result of the elimination method will be an equation that is never true, such as $0 = 17$. **(Solving systems of equations using the method of elimination)**

11. If the system of equations is dependent, the result of the elimination method will be an equation that is always true, like $0 = 0$. **(Solving systems of equations using the method of elimination)**

12. This system is perfectly set up for the elimination method. Simply add the two equations together.

$$\begin{array}{r} 2x + y = 5 \\ \underline{x - y = 1} \\ 3x = 6 \end{array}$$ Thus, $x = 2$.

Using this value in $2x + y = 5$, we find $2(2) + y = 5$. Thus, $y = 1$. The common solution is (2, 1). **(Solving systems of equations using the method of elimination)**

13. Multiply the second equation by –2 to eliminate the variable x.

$$2x + 3y = 7 \qquad \rightarrow \quad 2x + 3y = 7$$
$$-2(x + y) = (2)(-2) \rightarrow -2x - 2y = -4$$

Now, add the two equations together:

$$\begin{array}{r} 2x + 3y = 7 \\ \underline{-2x - 2y = -4} \\ y = 3 \end{array}$$

Using this value in $x + y = 2$, we obtain $x + 3 = 2$ or $x = -1$. Thus, the common solution is (–1,3). **(Solving systems of equations using the method of elimination)**

14. Multiply the first equation by 3 and the second equation by 2 to eliminate the variable y:

$$3(3x + 2y) = 8(3) \;\rightarrow\; 9x + 6y = 24$$
$$2(2x - 3y) = (14)(2) \;\rightarrow\; 4x - 6y = 28$$

Now, add the two equations together.

$$9x + 6y = 24$$
$$\underline{4x - 6y = 28}$$
$$13x = 52 \;\; \text{Thus, } x = 4.$$

Using this value in $3x + 2y = 8$, we obtain $3(4) + 2y = 8$ or $2y = -4$. Thus, $y = -2$, and the common solution is $(4, -2)$. **(Solving systems of equations using the method of elimination)**

15. Subtract the second equation from the first to eliminate the variable y:

$$3x + y = p$$
$$\underline{-(x + y) = -q}$$
$$2x = p - q \;\; \text{Thus, } x = \frac{(p - q)}{2}.$$

Using this value in $x + y = q$, we obtain $\dfrac{(p - q)}{2} + y = q$ or $y = q - \dfrac{(p - q)}{2}$.

The value for y can be simplified: $y = 2q - \dfrac{(p - q)}{2} = \dfrac{2q}{2} - \dfrac{p}{2} + \dfrac{q}{2} = \dfrac{3q}{2} - \dfrac{p}{2} = \dfrac{3q - p}{2}$.

Thus, the common solution is $\left(\dfrac{(p - q)}{2}, \dfrac{(3q - p)}{2}\right)$. **(Solving systems of equations using the method of elimination)**

16. Multiply the first equation by 2 to eliminate the variable y:

$$2(3x - 2y) = (10)2 \;\rightarrow\; 6x - 4y = 20$$
$$-6x + 4y = 8 \qquad\quad \rightarrow\; -6x + 4y = 8$$

Now, add the two equations together:

$$6x - 4y = 20$$
$$\underline{-6x + 4y = 8}$$
$$0 = 28$$

Since $0 = 28$ is an impossible situation, there is no common solution for the two equations. The system is inconsistent. Note, in fact, that this is exactly the same system we graphed in problem 8, obtaining two parallel lines, which also indicates no solution. **(Solving systems of equations using the method of elimination)**

17. To start, let's multiply the top equation by –2 to eliminate x:

$-2(2x - 8y) = -2(6) \rightarrow -4x + 16y = -12$
$4x - 16y = 12 \qquad \rightarrow \qquad 4x + 16y = 12$

Now, add the equations together:

$-4x + 16y = -12$
$\underline{4x + 16y = 12}$
$\qquad\quad 0 = 0$

Since $0 = 0$ is always true, we have a system that has an infinite number of solutions. In fact, when we graphed this system (problem 9), we found that both lines were the same. The system is dependent; any point that satisfies one equation satisfies both equations. (**Solving systems of equations using the method of elimination**)

18. If one of the equations in the system is already solved for one of the variables or if one of the equations in the system can easily be solved for one of the variables, then it is usually easier to use the substitution method. (**Solving systems of equations using substitution**)

19. This system is set up for an easy application of the substitution method. Since we are given $y = x - 2$, we simply need to substitute this value for y into the second equation, $2y = x - 3$.
$2y = x - 3$ tells us that $2(x - 2) = x - 3$. Now, we solve for x: $2(x - 2) = x - 3$. So $2x - 4 = x - 3$ or $x = 1$.

Thus, $y = x - 2 = 1 - 2 = -1$. The common solution is $(1, -1)$. (**Solving systems of equations using substitution**)

20. Substitute $y = 3x + 2$ into $x + y = 6$. We get $x + 3x + 2 = 6$, or $4x = 4$, so $x = 1$. Now, we know that $x + y = 6$, so, if $x = 1$, $y = 5$. The common solution is $(1, 5)$. (**Solving systems of equations using substitution**)

21. Rewrite $2x - y = 3$ as $y = 2x - 3$. Substitute this result into $3x + y = 22$. We get $3x + (2x - 3) = 22$, so that $5x = 25$ and $x = 5$. Since $y = 2x - 3$, we have $y = 2(5) - 3 = 10 - 3 = 7$. Thus, the common solution is $(5, 7)$. (**Solving systems of equations using substitution**)

22. Rewrite the second equation as $y = 7x - 6$. Substitute this result into $3x + y = 4$. We obtain $3x + (7x - 6) = 4$, or $10x - 6 = 4$. This means $10x = 10$, or $x = 1$. Since $y = 7x - 6$, we have $y = 7(1) - 6 = 7 - 6 = 1$. Thus, the common solution is $(1, 1)$. (**Solving systems of equations using substitution**)

23. To use the method of elimination with a system of three equations in three unknowns, we must begin by reducing the system to a 2×2 system. To do this, we take two of the equations at a time, and eliminate one variable. Let's start with the first two equations.

$x - 2y + 3z = 10$
$x + 2y - z = 6$

We can subtract the second equation from the first to eliminate x.

$\quad x - 2y + z = 10$
$\underline{-(x + 2y - z) = -(6)}$
$\qquad\quad -4y + 4z = 4$

If we divide this equation by 4, we simplify it to $-y + z = 1$.

Now, let's add twice the first equation to the third to eliminate x from the sum of these two equations.

$$
\begin{array}{rcl}
2(x - 2y + 3z) = (10)2 & \rightarrow & 2x - 4y + 6x = 20 \\
-2x + y + 2z = -2 & \rightarrow & \underline{-2x + y + 2z = -2} \\
& & -3y + 8z = 18
\end{array}
$$

We now have a new system of two equations in two unknowns, and we can solve it in the usual way. Multiply the top equation by -3 to eliminate y.

$$
\begin{array}{rcl}
-3(-y + z) = -3(1) & \rightarrow & 3y - 3z = -3 \\
-3y + 8z = 18 & \rightarrow & \underline{-3y + 8z = 18} \\
& & 5z = 15
\end{array}
$$

From the fact that $5z = 15$, we have $z = 3$. Then, since $-y + z = 1$, we have $-y + 3 = 1$, or $y = 2$. Now, let's go to one of the original equations: $x - 4 + 9 + 10$. With $y = 2$ and $z = 3$, we get $x - 2(2) + 3(3) = 10$. Thus, $x - 4 + 9 = 10$, and $x + 5 = 10$. Therefore, $x = 5$. Finally, the common solution is (5, 2, 3). (**Solving systems of equations using the method of elimination**)

24. In the system,
$x - y + z = 4$
$3x + y + 2z = 20$
$4x - y + 5z = 29$. Let's begin by adding the first two equations together to eliminate y.

$$
\begin{array}{r}
x - y + z = 4 \\
\underline{3x + y + 2z = 20} \\
4x + 3z = 24
\end{array}
$$

Now, let's add the second and third equations.

$$
\begin{array}{r}
3x + y + 2z = 20 \\
\underline{4x - y + 5z = 29} \\
7x + 7z = 49
\end{array}
$$
This equation can be simplified by dividing by 7, to get $x = z = 7$.

Therefore, we have reduced the problem to solving the system

$4x + 3z = 24$
$x + z = 7$

Now multiply the second equation by -3 and add.

$$
\begin{array}{rcl}
4x + 3z = 24 & \rightarrow & 4x + 3z = 24 \\
-3(x + z) = (7)(-3) & & \underline{-3x - 3z = -21} \\
& & x = 3
\end{array}
$$

Since $x = 3$, we know from the fact that $x + z = 7$, $z = 4$. Finally, since $x - y + z = 4$, we have $3 - y + 4 = 4$, or $y = 3$. Thus, the common solution is (3, 3, 4). (**Solving systems of equations using the method of elimination**)

25. Let x = the larger of the numbers. Let y = the smaller of the numbers. Then, we have

$x + y = 27$
$x - y = 11$

It seems as if the easiest way to solve this system is simply to add the two equations to eliminate y:

$x + y = 27$
$\underline{x - y = 11}$
$\quad 2x = 38$

Thus, $x = 19$, and y must be 8.
The two numbers are 19 and 8. **(Solving word problems using systems of equations)**

26. Let x = the length of the longer piece. Let y = the length of the shorter piece. Then:

$x = y = 24$
$x - y = 2$

For practice, let's solve this system by substitution. From the second equation, we have $x = y + 2$. Plugging into the first equation, we get $y + 2 + y = 24$, so $2y = 22$ and $y = 11$. If $y = 11$, then $x = 13$. The length of the shorter piece of wood is 11 feet. **(Solving word problems using systems of equations)**

27. Let x = the number of votes the winner received. Let y = the number of votes the loser got. Then, we have

$x + y = 10,764$
$x - y = 122$

Adding the equations together, we get $2x = 10,886$.
Thus, $x = 5,443$. And y, which is 122 less than x, is $5,443 - 122 = 5,321$.
The winning candidate got 5,443 votes; the opponent got 5,321. **(Solving word problems using systems of equations)**

28. Let D = the number of dimes. Let Q = the number of quarters. Then, we have

$D + Q = 34$
$10D + 25Q = 550$

This system seems to be easiest to solve by substitution. From the first equation, we have $D = 34 - Q$. Plugging this into the second equation,

$10(34 - Q) + 25Q = 550$ or
$340 - 10Q + 25Q = 550$ and
$15Q = 210$. Thus, $Q = 14$ and $D = 20$

Brian has 14 quarters and 20 dimes. **(Solving word problems using systems of equations)**

29. The determinant of a matrix that contains either a row or a column of 0's is always 0. **(Determinants)**

30. The determinant of a matrix that has two rows that are the same is always 0. **(Determinants)**

31. Let $\begin{vmatrix} a_1 & b_1 \\ a_2 & b_2 \end{vmatrix}$ represent a 2 × 2 determinant. Then, the value of the determinant is given by $a_1b_2 - a_2b_1$.
(Determinants)

32. By the formula in problem 31, the determinant $\begin{vmatrix} 10 & 7 \\ 2 & -7 \end{vmatrix}$ is equal to

    $(10)(-7) - (7)(2) = -70 - 14 = -84.$ **(Determinants)**

33. The determinant $\begin{vmatrix} -5 & 6 \\ -3 & 4 \end{vmatrix}$ is equal to $(-5)(4) - (6)(-3) = -20 + 18 = -2.$ **(Determinants)**

34. The determinant $\begin{vmatrix} 6 & 8 \\ 3 & 4 \end{vmatrix}$ is equal to $(6)(4) - (3)(8) = 24 - 24 = 0.$ **(Determinants)**

35. The determinant $\begin{vmatrix} 13 & 25 \\ 0 & 0 \end{vmatrix}$ is equal to $(13)(0) - (0)(25) = 0.$ **(Determinants)**

36. Let a system of two linear equations in two variables be given by
    $$a_1 x + b_1 y = c_1$$
    $$a_2 x + b_2 y = c_2$$

    Define the following three determinants:

    $$D = \begin{vmatrix} a_1 & b_1 \\ a_2 & b_2 \end{vmatrix} \quad D_x = \begin{vmatrix} c_1 & b_1 \\ c_2 & b_2 \end{vmatrix} \quad D_y = \begin{vmatrix} a_1 & c_1 \\ a_2 & c_2 \end{vmatrix}$$

    Then, if the common solution of the system is (x, y), we can compute the values of x and y as follows:

    $$x = \frac{D_x}{D} \quad \text{and} \quad y = \frac{D_y}{D}$$

    It is easy to prove that Cramer's Rule gives the correct solution in the 2×2 case. Once again, consider the system of equations:
    $$a_1 x + b_1 y = c_1$$
    $$a_2 x + b_2 y = c_2$$

    To find the common solution of this system, let's try to use the elimination method. Multiply the top equation by b_2, and the bottom equation by b_1 and then subtract the bottom equation from the top to eliminate y.
    $$a_1 b_2 x + b_1 b_2 y = c_1 b_2$$
    $$a_2 b_1 x + b_1 b_2 y = c_2 b_1$$
    $$a_1 b_2 x - a_2 b_1 x = c_1 b_2 - c_2 b_1 \quad \text{Factor } x \text{ from the left-hand side}$$
    $$x(a_1 b_2 - a_2 b_1) = c_1 b_2 - c_2 b_1$$

    Divide by $a_1 b_2 - a_2 b_1$:
    $$x = \frac{(c_1 b_2 - c_2 b_1)}{(a_1 b_2 - a_2 b_1)}$$

    Now, note that the value of D_x is exactly $c_1 b_2 - c_2 b_1$, and the value of D is $a_1 b_2 - a_2 b_1$. Thus, $x = \frac{Dx}{D}$.

    Similar logic can be used to prove that $y = \frac{Dy}{D}$. **(Solving systems of equations using Cramer's Rule)**

37. When using Cramer's Rule with a system of equations that is inconsistent, the determinant of the *D* matrix will be equal to 0, making it impossible to compute values for *x* and *y*. (**Solving systems of equations using Cramer's Rule**)

38. When using Cramer's Rule to solve a dependent system of equations, all three of the determinants that need to be computed will turn out to equal 0, indicating that there is no unique solution (*x*, *y*). (**Solving systems of equations using Cramer's Rule**)

39. For the system
$$2x - 3y = 12$$
$$4x + 5y = 2$$

we define the following three determinants:

$$D = \begin{vmatrix} 2 & -3 \\ 4 & 5 \end{vmatrix} \quad D_x = \begin{vmatrix} 12 & -3 \\ 2 & 5 \end{vmatrix} \quad D_y = \begin{vmatrix} 2 & 12 \\ 4 & 2 \end{vmatrix}$$

By Cramer's Rule, $x = \dfrac{D_x}{D}$ and $y = \dfrac{D_y}{D}$. Evaluating the determinants, we get

$$D = (2)(5) - (-3)(4) = 10 + 12 = 22; \quad D_x = (12)(5) - (-3)(2) = 60 + 6 = 66;$$
$$D_y = (2)(2) - (12)(4) = 4 - 48 = -44$$

Thus, $x = \dfrac{66}{22} = 3$, and $y = \dfrac{-44}{22} = -2$. The common solution is (3, –2), which is the same solution that we obtained in problem 6. (**Solving systems of equations using Cramer's Rule**)

40. For the system
$$4x + 3y = 15$$
$$-x + 4y = 20$$

we define the following three determinants:

$$D = \begin{vmatrix} 4 & 3 \\ -1 & 4 \end{vmatrix} \quad D_x = \begin{vmatrix} 15 & 3 \\ 20 & 4 \end{vmatrix} \quad D_y = \begin{vmatrix} 4 & 15 \\ -1 & 20 \end{vmatrix}$$

By Cramer's Rule, $x = \dfrac{D_x}{D}$ and $y = \dfrac{D_y}{D}$. Evaluating the determinants, we get

$$D = (4)(4) - (3)(-1) = 16 + 3 = 19$$
$$D_x = (15)(4) - (20)(3) = 60 - 60 = 0$$
$$D_y = (4)(20) - (15)(-1) = 80 + 15 = 95$$

Thus, $x = \dfrac{0}{19} = 0$, and $y = \dfrac{95}{19} = 5$. The common solution is (0, 5), which is the same solution that we obtained in problem 7. (**Solving systems of equations using Cramer's Rule**)

41. For the system
$$3x - 2y = 10$$
$$-6x + 4y = 8$$
we define the following three determinants:

$$D = \begin{vmatrix} 3 & -2 \\ -6 & 4 \end{vmatrix} \quad D_x = \begin{vmatrix} 10 & -2 \\ 8 & 4 \end{vmatrix} \quad D_y = \begin{vmatrix} 3 & 10 \\ -6 & 8 \end{vmatrix}$$

By Cramer's Rule, $x = \dfrac{D_x}{D}$ and $y = \dfrac{D_y}{D}$. Evaluating the determinants, we get

$$D = (3)(4) - (-2)(-6) = 12 - 12 = 0$$
$$D_x = (10)(4) - (-2)(8) = 40 + 16 = 56$$
$$D_y = (3)(8) - (10)(-6) = 24 + 60 = 84$$

Since D is 0, and division by 0 is undefined, we cannot define values for x and y. The system is inconsistent, which is exactly what we discovered in problem 8. **(Solving systems of equations using Cramer's Rule)**

42. For the system
$$2x - 8y = 6$$
$$4x - 16y = 12$$
we define the following three determinants:

$$D = \begin{vmatrix} 2 & -8 \\ 4 & -16 \end{vmatrix} \quad D_x = \begin{vmatrix} 6 & -8 \\ 12 & -16 \end{vmatrix} \quad D_y = \begin{vmatrix} 2 & 6 \\ 4 & 12 \end{vmatrix}$$

By Cramer's Rule, $x = \dfrac{D_x}{D}$ and $y = \dfrac{D_y}{D}$. Evaluating the determinants, we get

$$D = (2)(-16) - (-8)(4) = -32 + 32 = 0$$
$$D_x = (6)(-16) - (-8)(-12) = -96 + 96 = 0$$
$$Dy = (2)(12) - (6)(4) = 24 - 24 = 0$$

All three determinants are equal to 0; this means that the system is dependent. Note that we obtained the same result in problem 9. **(Solving systems of equations using Cramer's Rule)**

43. Before we can begin to solve this problem, we need to determine how to evaluate a 3×3 determinant. Consider the following determinant:

$$\begin{vmatrix} a_1 & b_1 & c_1 \\ a_2 & b_2 & c_2 \\ a_3 & b_3 & c_3 \end{vmatrix}$$

The value of this determinant is $a_1b_2c_3 + a_2b_3c_1 + a_3b_1c_2 - a_3b_2c_1 - a_2b_1c_3 - a_1b_3c_2$. There is an easy diagrammatic way to remember this formula. First of all, find the three products indicated by the arrows on the determinant. Find the sum of these three products.

Next, find the three products indicated by the arrows on the determinant below.

Subtract the sum of these three products from the sum obtained above. Thus, in the case of the determinant given for this problem, we have $(0)(5)(0) + (0)(4)(3) + (-2)(0)(0) - (3)(5)(-2) - (0)(0)(0) - (0)(4)(0) = 30$. **(Determinants)**

44. Using the formula developed in problem 43, we have $(5)(0)(1) + (-4)(-1)(0) + (1)(6)(2) - (0)(0)(1) - (6)(-4)(1) - (5)(-1)(2) = 12 + 24 + 10 = 46$. **(Determinants)**

45. Recall that any matrix with a row of 0's has a determinant of 0. In fact, if you substitute the values from the determinant into the formula, you will see that a 0 from the second row appears in every term, making the value of the determinant 0. **(Determinants)**

46. Note that the first and third rows of the matrix are the same. As already discussed, this means that the value of the determinant is 0. Of course, you could also do the computation to demonstrate that it is, in fact, 0. **(Determinants)**

47. Cramer's Rule for a system of three equations in three unknowns (3×3 system) is analogous to that for the 2×2 system. First of all, consider the 3×3 system given below:
$$a_1 x + b_1 y + c_1 z = d_1$$
$$a_2 x + b_2 y + c_2 z = d_2$$
$$a_3 x + b_3 y + c_3 z = d_3$$
For this system, define the following four determinants:

$$D = \begin{vmatrix} a_1 & b_1 & c_1 \\ a_2 & b_2 & c_2 \\ a_3 & b_3 & c_3 \end{vmatrix} \quad D_x = \begin{vmatrix} d_1 & b_1 & c_1 \\ d_2 & b_2 & c_2 \\ d_3 & b_3 & c_3 \end{vmatrix} \quad D_y = \begin{vmatrix} a_1 & d_1 & c_1 \\ a_2 & d_2 & c_2 \\ a_3 & d_3 & c_3 \end{vmatrix} \quad D_z = \begin{vmatrix} a_1 & b_1 & d_1 \\ a_2 & b_2 & d_2 \\ a_3 & b_3 & d_3 \end{vmatrix}$$

Then, $x = \dfrac{D_x}{D}$, $y = \dfrac{D_y}{D}$, and $z = \dfrac{D_z}{D}$. Notice how unwieldy it is to solve a 3×3 system with Cramer's Rule. We will develop an easier technique shortly. In the meantime, let's use Cramer's Rule.

$D = (1)(2)(2) + (1)(1)(3) + (-2)(-2)(-1) - (3)(2)(-2) - (-2)(1)(2) - (1)(1)(-1) = 4 + 3 - 4 + 12 + 4 + 1 = 20$

Similarly, $D_x = 40 + 18 - 4 + 12 + 24 + 10 = 100$. Thus, $x = \dfrac{D_x}{D} = \dfrac{100}{20} = 5$.

In the same way, we compute $D_y = 40$ and $D_z = 60$. Thus, $y = \dfrac{D_y}{D} = \dfrac{40}{20} = 2$ and $z = \dfrac{60}{20} = 3$. The solution is (5, 2, 3), which is the same as the solution we got in problem 23. **(Solving systems of equations using Cramer's Rule)**

48. Again, we need to compute the values of the four determinants, D, D_x, D_y, and D_z. Using the same formula as before, we obtain $D = (1)(1)(5) + (3)(-1)(1) + (4)(2)(-1) - (1)(1)(4) - (-1)(3)(5) - (1)(-1)(2) = 5 - 3 - 8 - 4 + 15 + 2 = 7$.

Similarly, $D_x = 20 - 20 - 58 - 29 + 100 + 8 = 21$. Thus, $x = \dfrac{D_x}{D} = \dfrac{21}{7} = 3$. In the same fashion, we compute $D_y = 21$ and $D_z = 28$. Therefore, we have $y = 3$ and $z = 4$. The common solution is $(3, 3, 4)$, which is the same solution we obtained in problem 24. (**Solving systems of equations using Cramer's Rule**)

49. A matrix is simply a rectangular array of numbers. While matrices have many uses in mathematics, in this chapter, we will be using them to help solve systems of equations. Consider, for example, the following system of equations:
$$2x - y + z = 3$$
$$-x + y + 2z = -5$$
$$2x - 2y + 3z = 3$$

Since the variables x, y, and z are basically just placeholders, we can economize on space by positioning the coefficients of x, y, and z in a 3×3 matrix as follows:

$$\begin{pmatrix} 2 & -1 & 1 \\ -1 & 1 & 2 \\ 2 & -2 & 3 \end{pmatrix}$$

(**Matrices and row operations**)

50. Note that the matrix in problem 49 does not represent the entire system of equations but only the left-hand side. In order to represent the entire system of equations by a matrix, we add one more column to the matrix, and place the constants from the right-hand side of the equations in this column. This new matrix is called an *augmented matrix*. If the original matrix is represented by M, the augmented matrix is called M^*. (**Matrices and row operations**)

51. The system of equations in problem 12 can be represented in an augmented matrix as follows:

$$M^* = \begin{pmatrix} 2 & 1 & | & 5 \\ 1 & -1 & | & 1 \end{pmatrix}$$ (**Matrices and row operations**)

52. The system of equations in problem 47 can be represented in an augmented matrix as follows:

$$M^* = \begin{pmatrix} 1 & -2 & 3 & | & 10 \\ 1 & 2 & -1 & | & 6 \\ -2 & 1 & 2 & | & -2 \end{pmatrix}$$ (**Matrices and row operations**)

53. The three elementary matrix row operations are:

   1. the interchange of any two rows

   2. the multiplication of any row by a nonzero constant

   3. the termwise addition of any multiple of one row to any other row

Performing any elementary row operation on the matrix of a system of linear equations will yield the matrix of an equivalent system of linear equations, that is, a system of equations that has the same common solution as the original system. (**Matrices and row operations**)

54. Two matrices are row equivalent if one can be obtained from the other by performing a finite number of elementary matrix row operations. Thus, two row equivalent matrices represent systems of equations with the same common solution. (**Matrices and row operations**)

55. In a square matrix (a matrix with the same number of rows as columns), the main diagonal contains the entries in the first row and first column, second row and second column, and so on down. The main diagonal is indicated in the matrix below:

$$\begin{vmatrix} a_1 & b_1 & c_1 \\ a_2 & b_2 & c_2 \\ a_3 & b_3 & c_3 \end{vmatrix}$$

A matrix is in echelon form if all of the entries below the main diagonal are 0's. We will use matrices to solve systems of equations in the following way: write an augmented matrix for the system, and then perform elementary row operations on the matrix until the portion of the matrix holding the coefficients is in echelon form. Once this has been done, as can be seen in the upcoming problems, it is fairly easy to solve the system. (**Matrices and row operations**)

56. The system of equations from problem 12 is
$2x + y = 5$
$x - y = 1$

When this is entered into an augmented matrix, it looks like this:

$$M^* = \begin{pmatrix} 2 & 1 & | & 5 \\ 1 & -1 & | & 1 \end{pmatrix}$$

To put this matrix in echelon form, multiply the top row by $-\frac{1}{2}$, and add the result to the second row. The matrix then becomes $M^* = \begin{pmatrix} 2 & 1 & | & 5 \\ 0 & -\frac{3}{2} & | & -\frac{3}{2} \end{pmatrix}$.

This system has the same common solution as the original system. Now, the second row of the equation says $\left(\frac{-3}{2}\right)y = \frac{-3}{2}$, which tells us that $y = 1$. Substituting this result into the top row (representing the first equation), gives us $2x + 1 = 5$, or $x = 2$. The common solution is, then, $(2,1)$, which is the same result we obtained when we solved problem 12. (**Using matrices to solve systems of equations**)

57. The system of equations from problem 16 is
$3x - 2y = 10$
$-6x + 4y = 8$

The augmented matrix for this system is: $M^* = \begin{pmatrix} 3 & -2 & | & 10 \\ -6 & 4 & | & 8 \end{pmatrix}$.

To put this matrix in echelon form, multiply the top row by 2, and add it to the bottom row. We obtain the augmented matrix $\begin{pmatrix} 3 & -2 & | & 10 \\ 0 & 0 & | & 28 \end{pmatrix}$.

The bottom row translates into the equation $0 = 28$, which, of course, has no solutions. Thus, the system is inconsistent, which is the same result we obtained when we solved problem 16. **(Using matrices to solve systems of equations)**

The bottom row translates into the equation $0 = 28$, which of course has no solutions. Thus, the system is inconsistent, which is the same result we obtained when we solv problem 16. **(Using matrices to solve systems of equations)**

58. The system of equations from problem 22 is
    $3x + y = 4$
    $7x - y = 6$

    The augmented matrix for the system is $M^* = \begin{pmatrix} 3 & 1 & | & 4 \\ 7 & -1 & | & 6 \end{pmatrix}$.

    To put the matrix in echelon form, multiply the top row by $-\dfrac{7}{3}$, and add the result to the bottom row. We obtain a new matrix:

    $$\begin{pmatrix} 3 & 1 & | & 4 \\ 0 & -10/3 & | & -10/3 \end{pmatrix}$$

    The bottom row translates into the equation $\dfrac{-10}{3}y = \dfrac{-10}{3}$, and it can be seen that $y = 1$. Plugging into the top equation, we get $3x + 1 = 4$, or $x = 1$. Thus, the common solution is $(1,1)$. This is the same answer that we obtained before. **(Using matrices to solve systems of equations)**

59. Using matrices to solve a 3×3 system of equations is essentially the same as solving a 2×2 system, although usually it takes an extra step to get the matrix in echelon form. The equations from problem 23 are
    $$x - 2y + 3z = 10$$
    $$x + 2y - z = 6$$
    $$-2x + y + 2z = -2$$

    This system can be represented in an augmented matrix as:

    $$M^* = \begin{pmatrix} 1 & -2 & 3 & | & 10 \\ 1 & 2 & -1 & | & 6 \\ -2 & 1 & 2 & | & -2 \end{pmatrix}$$

    To get this matrix in echelon form, begin by multiplying the first equation by -1 and adding it to the second equation. Then, multiply the first equation by 2, and add it to the third equation. We obtain a new matrix:

$$\begin{pmatrix} 1 & -2 & 3 & 10 \\ 0 & 4 & -4 & -4 \\ 0 & -3 & 8 & 18 \end{pmatrix}$$

Now, multiply the second row by $\dfrac{3}{4}$, and add the result to the third row to obtain

$$\begin{pmatrix} 1 & -2 & 3 & 10 \\ 0 & 4 & -4 & -4 \\ 0 & 0 & 5 & 15 \end{pmatrix}$$

The matrix is now in echelon form. The last row tells us that $5z = 15$, or $z = 3$. The second row tells us that $4y - 4z = -4$. Since $z = 3$, we have $4y - 12 = -4$, or $y = 2$.

Finally, from the first row, $x - 2y + 3z = 10$. Using $y = 2$ and $z = 3$, we obtain $x - 4 + 4 + 9 = 10$. This means that $x = 5$. Thus, the common solution is $(5, 2, 3)$, which is what we obtained for problem 23. **(Using matrices to solve systems of equations)**

60. Let $x =$ one of the numbers. Let $y =$ the other number. Then, we obtain the system:
$x - y = 67$
$xy = 3300$

Note that the second equation is not linear. In such situations, we use the substitution method to solve for x and y. The second equation tells us that $y = \dfrac{3300}{x}$. Plugging this into the first equation, we obtain

$x - \dfrac{3300}{x} = 67$. Multiply both sides by x to clear the fraction.

$$x\left[x - \frac{3300}{x}\right] = 67x \text{ or } x^2 - 3300 = 67x$$

We have ended up with a quadratic equation. Set it equal to 0 and factor:

$$x^2 - 67x - 3300 = 0$$

$(x + 33)(x - 100) = 0$. We now set each factor separately equal to 0 and $x = 33, 100$. Now, if $x = -33$, $y = -100$. Further, if $x = 100$, $y = 33$. Thus, there are really two solutions to the problem. The two numbers are either 33 and 100, or -100 and -33. **(Solving word problems using systems of equations)**

61. Let $x =$ the amount invested at 7%.

Let $y =$ the amount invested at 8%.

Then, we have the following system of equations:
$x + y = 2000$
$.07x + .08y = 151$

First of all, multiply the second equation by 100 to clear the decimals. The system becomes
$x + y = 2000$
$7x + 8y = 15100$

Let's solve this system with the elimination method.
Multiply the top equation by -7.

$-7x - 7y = -14000$
$\underline{7x + 8y = 15100}$
$y = 1100$

This means that $x = 900$.

Thus, $900 was invested at 7%, and $1100 was invested at 8%. **(Solving word problems using systems of equations)**

62. Let $x =$ one of the numbers. Let $y =$ the other number.

Then, we have the following system:
$x = 8 + 4y$
$x + y = -2$

This system is set up for an easy application of the substitution method. Substitute $x = 8 + 4y$ into the second equation. Then, we get $8 + 4y + y = -2$, or $5y = -10$, so $y = -2$. If $y = -2$, then x = 0.

The common solution is $(0, -2)$. Thus, the two numbers are 0 and -2. **(Solving word problems using systems of equations)**

63. Let $J =$ John's age. Let $D =$ Dennis's age.

Then, we have the following system:
$J = 3 + D$
$J - 4 = 2(D - 4)$
The system can be rewritten as
$J - D = 3$
$J - 2D = -4$

Let's solve this system by Cramer's Rule. The determinant that gives the denominator of the solutions is

$$\begin{vmatrix} 1 & -1 \\ 1 & -2 \end{vmatrix}$$

The determinant of this matrix evaluates to $(1)(-2) - (-1)(1) = 2 + 1 = 1$. The matrix that gives us the numerator for the value of J is

$$\begin{vmatrix} 3 & -1 \\ -4 & -2 \end{vmatrix}$$

and its determinant has a value of $(3)(2) - (-1)(4) = 6 - 4 = -10$. The matrix that gives the numerator for the value of D is

$$\begin{vmatrix} 1 & 3 \\ 1 & -4 \end{vmatrix}$$

and its determinant has a value of $(1)(-4) - (3)(1) = -4 - 3 = -7$. Thus, we have $J = \dfrac{-10}{-1} = 10$, and

$$D = \dfrac{-7}{-1} = 7.$$

Therefore, John is 10 years old, and Dennis is 7. **(Solving word problems using systems of equations)**

64. Let A = the number of Type A printers produced. Let B = the number of Type B printers produced. Then, we have the following system:

$A + B = 55$
$100A + 150B = 7000$

This time, we'll use matrices to solve the system. The augmented matrix for the system is

$$M^* = \begin{pmatrix} 1 & 1 & 55 \\ 100 & 150 & 7000 \end{pmatrix}$$

To put this matrix in echelon form, multiply the top row by -100, and add the result to the bottom row. We obtain the new matrix

$$\begin{pmatrix} 1 & 1 & 55 \\ 0 & 50 & 1500 \end{pmatrix}$$

From the bottom row, we have $50B = 1500$, or $B = 30$. Then, A must be 25.

The company made 25 Type A printers, and 30 Type B printers. **(Solving word problems using systems of equations)**

Grade Yourself

Circle the numbers of the questions you missed, then fill in the total incorrect for each topic. If you answered more than three questions incorrectly, you need to focus on that topic. (If a topic has fewer than three questions and you had at least one wrong, we suggest you study that topic also. Read your textbook or a review book, or ask your teacher for help.)

Subject: Systems of Equations

Topic	Question Numbers	Number Incorrect
Checking the solution of a system of equations	1, 2	
Solving systems of equations—theory	3, 4, 5	
Solving systems of equations by graphing	6, 7, 8, 9	
Solving systems of equations using the method of elimination	10, 11, 12, 13, 14, 15, 16, 17, 23, 24	
Solving systems of equations using substitution	18, 19, 20, 21, 22	
Solving word problems using systems of equations	25, 26, 27, 28, 60, 61, 62, 63, 64	
Determinants	29, 30, 31, 32, 33, 34, 35, 43, 44, 45, 46	
Solving systems of equations using Cramer's Rule	36, 37, 38, 39, 40, 41, 42, 47, 48	
Matrices and row operations	49, 50, 51, 52, 53, 54, 55	
Using matrices to solve systems of equations	56, 57, 58, 59	

Sequences and Series

8

Brief Yourself

Recursive and Explicit Definitions of Sequences

A *sequence* is a function from the positive integers to the real numbers. A sequence can be defined in one of two ways. It can be given an *explicit definition*, which is a formula that gives the value of the *n*th term of the series as a function of *n*. For example,

$$f(n) = n^2$$

is an explicit definition of the *sequence of squares* whose first few elements are 1, 4, 9, 16, ...

Or a sequence can be given a *recursive definition*. In this form of definition, you specify the value of $f(1)$ and then define $f(k)$ in terms of $f(k-1)$. For example,

$$f(1) = 1$$

$$f(k) = f(k-1) + 2k - 1$$

is a recursive definition of the function whose first few elements are 1, 4, 9, 16, … the same sequence for which we gave an explicit definition above.

Arithmetic and Geometric Sequences

An *arithmetic sequence* is a sequence in which the difference between successive terms is a constant. The explicit definition of an arithmetic sequence has the form $f(n) = a + (n-1)d$, where *d* is called the *common difference* of the sequence. The recursive form of an arithmetic sequence has the form $f(1) = a, f(k) = f(k-1) + d$.

A *geometric sequence* is a sequence in which the ratio of successive terms is a constant. The explicit definition of a geometric sequence always has the form $f(n) = ar^{(n-1)}$, where *r* is called the common ratio of the sequence. The recursive form of definition of a geometric sequence has the form $f(1) = a$, $f(k) = rf(k-1)$.

Other Sequences

Many sequences are neither arithmetic nor geometric. One example is the *Fibonacci sequence*, which has recursive definition $f(1) = 1, f(2) = 1, f(k) = f(k-1) + f(k-2)$. Another is the *harmonic sequence*, which has the explicit definition $f(n) = \dfrac{1}{a + (n-1)d}$. There is also the *factorial sequence*, which has explicit definition $f(n) = n!$ and recursive definition $f(1) = 1, f(k) = kf(k-1)$. And another example is the sequence of squares, defined above.

Mathematical Induction

An important tool in proving facts about sequences is the technique called *mathematical induction*. To use this technique of proof, you prove that a formula is true for all positive n by first showing that it is true for $k = 1$ and then showing that if the formula is true for $n = k = 1$, then it is true for k. For example, let us prove that the sum $S(k)$ of the first k terms of the arithmetic sequence defined by $f(n) = 1$ is $\dfrac{n(n+1)}{2}$. Clearly $S(1) = \dfrac{1(2)}{2} = 1$, so the formula is valid for $k = 1$. Now assume the formula is true for $n = k - 1$. Then

$$S(k) = 1 + 2 + 3 + \ldots + (k-1) + k = [1 + 2 + \ldots + (k-1)] + k = (k-1)\frac{k}{2} + k =$$

$$\frac{k(k-1)}{2} + \frac{2k}{2} = \frac{k^2 - k + 2k}{2} = \frac{k^2 + k}{21} = \frac{k(k+1)}{2}$$

So we have shown that if the formula is true for $k - 1$, then it is true for k. Therefore the formula is true for all positive integers.

Finite Series

A *finite series* is defined as the sum of a finite number of terms of a sequence. The quantity $S(k)$, which we defined above as the sum of the first k terms of the sequence $f(n) = n$, is one very simple example of a series.

The sum of the first k terms of the arithmetic sequence defined by $f(n) = a + (n-1)d$ is $S(k) = ak + d(k-1)\dfrac{k}{2}$. This general formula can be proved by mathematical induction.

The sum of the first k terms of the geometric sequence defined by $f(n) = ar^{(n-1)}$ is $S(k) = \dfrac{a\left[r^k - 1\right]}{r - 1}$. This sum is easy to prove either by mathematical induction or by multiplying $S(k)$ by r and then showing that $rS(k) - S(k) = a[r^k - 1]$.

We use the notation $\displaystyle\sum_{i=1}^{k} = f(i)$ to denote the series defined by summing the first k terms of the sequence $f(i)$. The letter Σ is a capital Greek letter called *sigma*.

Infinite Series

Consider the sequence $S(k)$ consisting of the sum of the first k terms of a sequence defined by $f(n)$. It may happen that as k gets larger and larger the quantity $S(k)$ gets closer and closer to some definite and finite value S. If this happens, then we define this value to be the sum of the infinite series $\sum_{i=1}^{\infty} = f(i)$. We sometimes write this as $S(k) \to S$, or $\lim_{k \to \infty} S(k) = S$. In words, we say "$S(k)$ converges to S."

For example, consider the geometric sequence defined explicitly by $f(n) = \dfrac{1}{2^{(n-1)}}$ and recursively by $f(1) = 1, f(k) = \dfrac{f(k-1)}{2}$. Then by the formula given above we know that the sum of the first k terms, $S(k)$, is $\dfrac{\left(\frac{1}{2}\right)^{k-1}}{\left(\frac{1}{2}\right)-1} = 2 - \left(\frac{1}{2}\right)^{(k-1)}$. Clearly, as k gets larger and larger the second term in this expression gets closer and closer to 0, so $S(k)$ converges to 2. Informally we write

$$2 = 1 + \frac{1}{2} + \frac{1}{4} + \frac{1}{8} + \frac{1}{16} + \frac{1}{32} + \dots$$

Infinite series are very important in higher mathematics because there are many, many functions that can be expressed as the sum of infinite series. To derive these series, it is best to use calculus, but here are some example of functions that can be expressed as infinite series:

$$\exp(x) = 1 + x + \frac{x^2}{2!} + \frac{x^3}{3!} + \frac{x^4}{4!} + \dots$$

$$\sin(x) = x - \frac{x^3}{3!} + \frac{x^5}{5!} - \frac{x^7}{7!} + \dots \text{ (x in radians)}$$

$$\cos(x) = 1 - \frac{x^2}{2!} + \frac{x^4}{4!} - \frac{x^6}{6!} + \dots \text{ (x in radians)}$$

The first function, $\exp(x)$, is simply e (which is 2.718281828...) raised to the power x. All three of these series are convergent for all values of x.

Test Yourself

1. Give an explicit and a recursive definition of the sequence 1, 2, 4, 8, 16, ….

2. Use mathematical induction to prove that your explicit and recursive definitions for problem 1 define the same sequence.

3. Give an explicit and a recursive definition for the sequence 9, 3, 1, $\frac{1}{3}$, … What is the tenth element in this sequence?

4. Give an explicit definition for the sequence 2, 5, 8, 11, 14, … List the next three elements in this sequence.

5. Give a recursive definition for the sequence 0, 2, 5, 9, 14, 20, … List its next three elements.

6. List the first five elements of the sequence defined by the recursive definition $f(1) = 1$, $f(k) = kf(k − 1)$ for $k > 1$. Give an explicit definition for this sequence.

7. Write out the first few terms of the sequence defined by $f(1) = 1$, $f(k) = 2f(k − 1) + 1$ for $k > 1$. Give an explicit definition of the sequence.

8. Use mathematical induction to show that the sum of the cubes of three consecutive integers is always divisible by 9.

Identify whether the sequences in problems 9–18 are arithmetic, geometric, or neither.

9. 2, 4, 6, 8, …

10. 1, 2, 4, 8, …

11. 1, 2, 3, 5, 7, …

12. 4, 2, 1, $\frac{1}{2}$, …

13. 19, 9, −1, −11, …

14. 2, 8, 26, 80, …

15. 5, 11, 17, 23, 29, …

16. $\sqrt{2}$, 2, $2\sqrt{2}$, 4, …

17. 3, −6, 12, −24, …

18. 1, 4, 9, 16, 25, …

Give an explicit definition for the sequences in problems 19–24.

19. 5, 8, 11, 14, …

20. 4, −2, 1, $−\frac{1}{2}$, $\frac{1}{4}$, …

21. 20, 30, 45, 67.5, …

22. 6, 7, 9, 13, 21, 37, 69, …

23. 1, 2, 6, 24, 120, 720, …

24. 1, 4, 9, 16, 25, 36, 49, …

25. If 2 and 4 are the first two elements of an arithmetic sequence, what is the explicit definition of the sequence? If 2 and 4 are the first two elements of a geometric sequence, what is the explicit definition of the sequence?

26. If 1, 4, 6 + x are the first three terms of a geometric sequence, what is the value of x? What is the value of x if it is an arithmetic sequence?

27. Use mathematical induction to establish the following formula:
$1 + 8 + 16 + … + 8(n − 1) = (2n − 1)^2$

28. Prove by mathematical induction that the sum of the first n terms of the sequence $f(n) = ar^{(n − 1)}$ is $\dfrac{a[r^n − 1]}{r − 1}$.

29. Prove by mathematical induction that the sum of the first n terms of the sequence $f(n) = n^2$ is $\dfrac{n(n + 1)(2n + 1)}{6}$.

30. Find the sum of the first ten terms of the sequence defined by $f(n) = 2 + 3n$.

31. Find the sum of the first five terms in the sequence defined by $f(n) = 3(2^{(n-1)})$.

32. Find the sum of the first five terms in the sequence defined by $f(n) = 5\left(\dfrac{1}{2}\right)^{(n-1)}$. Find the sum of the first 10 terms in the same sequence. What do you think the sum of the first 1,000 terms in the sequence would be (to the nearest .0001)?

33. Let $S(n)$ denote the sum of the first n terms in the sequence defined by $f(n) = n2^n$. Write the first four terms of the sequence, and calculate the values of $S(1)$, $S(2)$, $S(3)$, and $S(4)$. Prove by mathematical induction that the sum of the first n terms of the sequence is $2[(n-1)2n + 1]$.

For each of the series in problems 34–42, state whether it converges or does not converge. If it converges, give its value.

34. $4 - 2 + 1 - \dfrac{1}{2} \dots$

35. $1 + 1.5 + 2 + 2.5 \dots$

36. $10 + 9 + 8.1 + 7.29 + \dots$

37. $1 - 1.5 + 2 - 2.5 + 3.0 - \dots$

38. $1 + 1 + \dfrac{1}{2!} + \dfrac{1}{3!} + \dfrac{1}{4!} + \dfrac{1}{5!} + \dots$

39. $1 - \dfrac{1}{3!} + \dfrac{1}{5!} - \dfrac{1}{7!} \dots$

40. $1 - \dfrac{1}{2!} + \dfrac{1}{4!} - \dfrac{1}{6!} + \dots$

41. $1 - 1 + \dfrac{1}{2!} - \dfrac{1}{3!} + \dfrac{1}{4!} \dots$

42. $21 - 7 + \dfrac{7}{3} - \dfrac{7}{9} + \dots$

43. Use a power series to calculate sin(.1) to four decimal places (the angle measured in radians.)

44. Use a power series to calculate cos(.1) to four decimal places.

45. Use the results of questions 43 and 44 to calculate sin(.2) to four decimal places.

46. Find the common ratio of an infinite geometric series with sum 8 and first term 4.

47. Find the first three terms of an infinite geometric series with sum 81 and common ratio $\dfrac{1}{3}$.

48. For what value of x does the series $1 + 2x + 4x^2 + \dots$ converge to $\dfrac{3}{5}$?

49. Consider an infinite geometric series with positive terms and a common ratio r which has a finite limit S. What will the sum of the odd terms of the series be, and what will be the sum of the even terms?

Check Yourself

1. $f(n) = 2^{(n-1)}$ is the explicit formula, and $f(1) = 1$, $f(k) = 2f(k-1)$ if $k > 1$ is the recursive one. (**Definitions of sequences**)

2. $f(1) = 1 = 2^{(0)}$, so the definitions agree for $n = 1$. Let us assume they agree for $k - 1$. Then $f(k) = 2f(k-1) = 2[2^{(k-2)}] = 2^{(k-1)}$, so they agree for k. Thus, they agree for all positive integers n. (**Definitions of sequences**)

3. Explicit definition: $f(n) = 9[3^{(1-n)}]$. Recursive definition: $f(1) = 9$, $f(k) = \dfrac{f(k-1)}{3}$ if $k > 1$. The 10th element would be, from the explicit definition, $9[3^{(-9)}] = 3^{(-7)} = \dfrac{1}{2187}$. (**Definitions of sequences**)

4. $f(n) = 3n - 1$. The next three elements are 17, 20, 23. (**Definitions of sequences**)

5. $f(1) = 0$, $f(k) = f(k-1) + k$ for $k > 1$. The next three elements in the sequence are 27, 35, 44. (**Definitions of sequences**)

6. The first five elements of the sequence are 1, 2, 6, 24, 120, ... and the explicit formula for the sequence is $f(n) = n!$ (that is, n factorial, the product of the first n positive integers). (**Definitions of sequences**)

7. The first six terms of the sequence are 1, 3, 7, 15, 31, 63, ... and the explicit definition of the sequence is $f(n) = 2^n - 1$. (**Definitions of sequences**)

8. $1^3 + 2^3 + 3^3 = 1 + 8 + 27 = 36 = 4(9)$, so the statement is true for $n = 1$. Suppose that $(k-1)^3 + k^3 + (k+1)^3 = 9m$. Then $k^3 + (k+1)^3 + (k+2)^3 = 9m + (k+2)^3 - (k-1)^3 = 9m + 9(k^2 + k + 1) = 9[m + k^2 + k + 1]$, so if the statement holds for $k - 1$, it holds for k. This proves that it is true for all positive integers n. (**Mathematical induction**)

9. 2, 4, 6, 8 are the first four terms of an arithmetic sequence, with common difference 2. (**Arithmetic and geometric sequences**)

10. 1, 2, 4, 8 are the first four terms of a geometric sequence, with common ratio 2. (**Arithmetic and geometric sequences**)

11. 1, 2, 3, 5, 7 are the first five terms of neither an arithmetic nor a geometric sequence. (**Arithmetic and geometric sequences**)

12. 4, 2, 1, $\dfrac{1}{2}$ are the first four terms of a geometric sequence with common ratio $\dfrac{1}{2}$. (**Arithmetic and geometric sequences**)

13. 19, 9, −1, −11 are the first four terms of an arithmetic sequence with common difference −10. (**Arithmetic and geometric sequences**)

14. 2, 8, 26, 80 are the first four terms of neither an arithmetic nor a geometric sequence. (**Arithmetic and geometric sequences**)

15. 5, 11, 17, 23, 29 are the first five terms of an arithmetic sequence, with common difference 6. **(Arithmetic and geometric sequences)**

16. $\sqrt{2}, 2, 2\sqrt{2}, 4$ are the first four terms of a geometric sequence, with common ratio $\sqrt{2}$. **(Arithmetic and geometric sequences)**

17. 3, –6, 12, –24 are the first four terms of a geometric sequence, with common ratio –2. **(Arithmetic and geometric sequences)**

18. 1, 4, 9, 16, 25 are the first five terms of neither an arithmetic nor a geometric sequence. **(Arithmetic and geometric sequences)**

19. $f(n) = 2 + 3n$. **(Definitions of sequences)**

20. $f(n) = (-8)\left(-\dfrac{1}{2}\right)^n$. **(Definitions of sequences)**

21. $f(n) = 20(1.5)^{(n-1)}$. **(Definitions of sequences)**

22. $f(n) = 5 + (2)^{(n-1)}$. **(Definitions of sequences)**

23. $f(n) = n!$. **(Definitions of sequences)**

24. $f(n) = n^2$. **(Definitions of sequences)**

25. $f(n) = 2n$ is the explicit formula if it is an arithmetic sequence, and $f(n) = 2^n$ is the explicit formula if it is a geometric sequence. **(Arithmetic and geometric sequences)**

26. If the sequence is geometric, x has the value 10. If the series is arithmetic, x has the value 1. **(Arithmetic and geometric sequences)**

27. The formula obviously holds for $n = 1$. Assume it holds for $n = k - 1$, and try to show it is true for $n = k$. The sum of the first k terms will be
$(2(k-1)-1)^2 + 8(k-1) = (2k-3)^2 + 8k - 8 = 4k^2 - 12k + 9 + 8k - 8 = 4k^2 - 4k + 1 = (2k-1)^2$.
(Mathematical induction)

28. The formula obviously holds for $n = 1$. Assume it holds for all k less than n, where $n \neq 1$. Then the sum of the first n terms will be
$$ar^{(n-1)} + a\frac{\left[r^{(n-1)} - 1\right]}{r-1} = \frac{\left[ar^{(n-1)}(r-1) + ar^{(n-1)} - a\right]}{r-1} = \frac{a\left[r^n - r^{(n-1)} + r^{(n-1)} - 1\right]}{r-1} = \frac{a\left[r^n - 1\right]}{r-1}.$$
(Mathematical induction)

29. The formula holds for $n = 1$, since $\dfrac{1(1+1)(2+1)}{6} = \dfrac{1(1+1)(2+1)}{6} = 1$. Assume it holds for all $k < n$. Then
the sum of the first n terms will be $n^2 + \dfrac{[(n-1)(n)(2n-1)]}{6} = \dfrac{\left[6n^2 + 2n^3 - 3n^2 + n\right]}{6} = \dfrac{n(n+1)(2n+1)}{6}$.
(Mathematical induction)

30. The first term is 5, the tenth term is 32, and there are ten terms all together, so the sum is

$$S(10) = 5 \cdot 10 + \frac{3(9)(10)}{2} = 50 + 3(9)(5) = 50 + 135 = 185. \textbf{ (Finite series)}$$

31. Applying the formula for the sum of a geometric series we have $\dfrac{3\left[2^5 - 1\right]}{[2 - 1]} = 93.$ **(Finite series)**

32. The first five terms are 5, 2.5, 1.25, .625, and .3125, which add up to 9.6875. Using the formula for the

sum of the first ten terms gives us $\dfrac{5\left[\left(\frac{1}{2}\right)^{10} - 1\right]}{\left\{\frac{1}{2} - 1\right\}} = 10[1 - (.5)^{10}] = 9.990234....$ Apparently, the sum of the

first thousand terms would be 10 (to the nearest .0001). **(Finite series)**

33. 2, 8, 24, 64 are the first four terms of the sequence, and the sums are 2, 10, 34, 98. The formula obviously holds for $n = 1$. Assume it holds for all k less than n, where $n \neq 1$. Then, the sum of the first n terms will be $n2^n + 2[(n-2)2^{n-1} + 1] = n2^n + 2[n2^{n-1} - 2^n + 1] = n2^n + n2^n - 2^{n+1} + 2 = 2n2^n - 2^{n+1} + 2 = 2[n2^n - 2^n + 1] = 2[(n-1)2^n + 1].$ **(Mathematical induction)**

34. The underlying sequence is $f(n) = 4\left(\dfrac{-1}{2}\right)^{(n-1)}$, so $S(n) = 4\left(\dfrac{\left(\frac{-1}{2}\right)^{n=1} - 1}{\frac{-1}{2} - 1}\right)$; as n becomes larger, the sum

approaches $4\left(\dfrac{-1}{\frac{-3}{2}}\right) = \dfrac{8}{3}.$ **(Infinite series)**

35. This series does not converge. **(Infinite series)**

36. This series is defined by the sequence $f(n) = 10(.9)^{(n-1)}$, so $S(n) = \dfrac{10\left[(.9)^n - 1\right]}{(.9 - 1)} = 100[1 - (.9)^n]$, which converges to 100. **(Infinite series)**

37. This series does not converge. **(Infinite series)**

38. This series is $\exp(1) = e = 2.71828...$ **(Infinite series)**

39. This series is $\sin(1) = .8415.$ **(Infinite series)**

40. This series is $\cos(1) = .5403.$ **(Infinite series)**

41. This series is $\exp(-1) = \dfrac{1}{e} = .3679.$ **(Infinite series)**

42. This geometric series is based on the sequence defined by $f(n) = 21(-\frac{1}{3})^{(n-1)}$, so its partial sum is

$$S(n) = \frac{21\left[\left(\frac{-1}{3}\right)^n - 1\right]}{\frac{-1}{3} - 1} = \frac{63\left[1 - \left(\frac{-1}{3}\right)^n\right]}{4}, \text{ which converges to } \frac{63}{4} = 15.75. \textbf{ (Infinite series)}$$

43. $\sin(.1) = .1 - \frac{.001}{6} + \frac{.00001}{120} = .0998.$ **(Infinite series)**

44. $\cos(.1) = 1 - \frac{.01}{2} + \frac{.0001}{24} = .9950.$ **(Infinite series)**

45. $\sin(.2) = 2\sin(.1)\cos(.1) = 2(.0998)(.9950) = .1986.$ **(Infinite series)**

46. You want $8 = S = \frac{a}{1-r} = \frac{4}{1-r}$, so $r = \frac{1}{2}.$ **(Infinite series)**

47. Set $S = \frac{a}{1-r} = 81$. It is given that $r = \frac{1}{3}$, so $\frac{3a}{2} = 81$, so $a = 54$. That means the first three terms of the series are 54, 18, 6. **(Infinite series)**

48. The first term is 1 and the common ratio is $2x$, so $S(n) = \left(\frac{(2x)^n - 1}{2x - 1}\right)$. That gives $S = \frac{1}{1 - 2x}$. Set $\frac{1}{1 - 2x} = \frac{3}{5}$ and you find $5 = 3 - 6x$, so $x = -\frac{1}{3}$. The first few terms of the series are $1 - \frac{2}{3} + \frac{4}{9} - \frac{8}{27} + \frac{16}{81}\cdots$ **(Infinite series)**

49. We have $\frac{a}{1-r} = S$. The sum of the odd terms of the series will be based on the sequence $f(n) = a(r^2)^{(n-1)}$, which gives a sum $\frac{a}{1-r^2} = \frac{S}{1+r}$. The sum of the even terms will therefore be $S\left[1 - \left(\frac{1}{1+r}\right)\right] = S\left[\frac{r}{1+r}\right]$. **(Infinite series)**

Grade Yourself

Circle the numbers of the questions you missed, then fill in the total incorrect for each topic. If you answered more than three questions incorrectly, you need to focus on that topic. (If a topic has fewer than three questions and you had at least one wrong, we suggest you study that topic also. Read your textbook or a review book, or ask your teacher for help.)

Subject: Sequences and Series

Topic	Question Numbers	Number Incorrect
Definitions of sequences	1, 2, 3, 4, 5, 6, 7, 19, 20, 21, 22, 23, 24	
Mathematical induction	8, 27, 28, 29, 33	
Arithmetic and geometric sequences	9, 10, 11, 12, 13, 14, 15, 16, 17, 18, 25, 26	
Finite series	30, 31, 32	
Infinite series	34, 35, 36, 37, 38, 39, 40, 41, 42, 43, 44, 45, 46, 47, 48, 49	

S0-BFD-445

Disney's
THE ARISTOCATS

MOUSE WORKS

J
Picture Bk
Disney

Not very long ago, at the turn of the century, there were no automobiles on the streets of Paris. Pedestrians took leisurely walks through the city, and the very rich, elegantly clad in the latest fashions, went out in their sumptuous carriages for promenades.

Madame Adelaide Bonfamille was a charming elderly lady, and she was very rich. She often took her four cats out in her carriage for fresh air and a tour of the city. Madame was so fond of her cats that she thought of them as her children.

Duchess was a beautiful white angora cat, and Berlioz, Toulouse, and Marie were her three very talented kittens. Berlioz aspired to be a composer; Toulouse, a famous painter; and Marie, a prima donna. They all loved Madame dearly. She was a very kind woman.

Madame Bonfamille was very proud of her adopted
children, but Edgar, her butler, was not so fond of them.

"Now look at that Berlioz!" he muttered under his
breath when they reached home that morning. "He
jumps on top of Frou-Frou the mare and Madame does
not say a word about it. She spoils them!"

Madame encouraged the kittens to pursue their talents,
even when they turned her home all topsy-turvy.

As soon as he entered the house, Berlioz made a dash for Madame's ball of red wool. *Swish!* It unravelled across the room.

"My little darlings have so much fun!" Madame thought. She looked over to Toulouse who was at his easel starting a painting. "Good work, Toulouse!" she said to the proud kitten. When Roquefort the mouse came out of his hole to play and grabbed hold of Berlioz's tail for a ride, she smiled. "Hello, Roquefort! Are you out for a walk?" she said.

Madame liked Roquefort. He was a very friendly mouse. Roquefort loved Madame and the cats because they never tried to chase him away.

Everyone was very happy in Madame Bonfamille's home except the grumpy Edgar. He was perhaps over-proud of his position as butler, and felt it was beneath him to play babysitter for four spoiled and very silly cats. "I am not a butler," he would often think to himself, "but a catsitter for four animals who think they are Aristocats!"

Berlioz soon lost interest in the ball of red wool and went to the piano, his favorite musical instrument.

Marie joined him, and to his virtuoso playing she sang, "Do, re, mi, fa, sol, la, ti, do! Ow, ow, ow, ow, ow, ow, ow, meow!"

Duchess sat on a velvet-cushioned armchair nearby and listened to her daughter sing and her son play.

"Marie will become a famous prima donna," she thought with a smile. "And Berlioz will give piano recitals all over the world!" Duchess was very grateful to Madame Bonfamille for her kindness and generosity. "We will always live here with her," she thought happily.

All of a sudden, the doorbell rang and startled
everyone. Berlioz, Toulouse, and Marie were
Aristocats, but they were also very curious kittens.
They ran to the front door to see who had come.

It was Mr. Georges Hautecourt, Madame
Bonfamille's attorney. Berlioz waved his paw at the
old gentleman. The kittens were very fond of him.

"Madame is upstairs in her salon, sir," said Edgar.
"Please follow me."

But Mr. Hautecourt was very old and suffered from arthritis. The old man could not climb the stairs. Edgar had to carry him on his back to Madame's salon!

"Oh, thank you. Thank you so very much," said the attorney when they reached the salon. When the door closed behind the old man, Edgar almost collapsed!

"What an exhausting man!" he muttered.

13

14

Madame, with her precious Duchess in her arms, welcomed her life-long friend.

"My dear, dear friend," she said, "I am so grateful that you should come at such short notice."

"It is my pleasure, my dearest Adelaide!" he said respectfully. He then stood up as straight as he could and bent over to kiss Madame's hand. Instead, he took hold of Duchess' furry white tail and kissed that! Madame Bonfamille giggled at her friend's mistake. Mr. Hautecourt turned red.

Meanwhile, Edgar had slipped into the room next door. The clever butler had guessed that Mr. Hautecourt's unexpected visit had to do with money... Madame's very large fortune. Edgar carefully listened to their conversation through the pipes. Madame Bonfamille did not have an heir. Edgar hoped that she would reward him in her will for his many years of loyalty and good service. He hoped to inherit her money.

"My dear Mr. Hautecourt," he overheard her tell the old man, "I have carefully thought about my situation and have found a solution which would make me very happy. I would like to leave my home and my fortune to my darling cats! They have made me so happy, and I am sure Edgar will be glad to look after them when I am gone."

Edgar thought that he would faint. The color drained from his face and his smile vanished. "She will leave everything to those cats and not a thing to me!" he thought with fury. "I won't have it! The cats must go!"

Edgar quickly went to the kitchen. Carefully he prepared the cats' dinner. He mixed their milk with breadcrumbs and a little honey and then added a few sleeping pills to the mixture.

"They will fall asleep and then I will take them far away and leave them there," he thought to himself with pleasure. "And Madame Bonfamille will have to leave her fortune to me – and only me!"

"Dinner's ready!" called Edgar to the cats. They all came running, and very soon the four hungry cats were lapping up their meal. Roquefort joined them and Duchess let him share her food. Edgar always prepared good dinners.

"Isn't this delicious!" asked Edgar with a crooked smile. "Edgar's special recipe for cats! You won't find anything like it in all of Paris."

As soon as the cats finished eating they started to feel very sleepy. They slowly crawled to their basket. Roquefort returned to his hole, where he fell sound asleep.

"They are exhausted!" thought Madame Bonfamille when she found her cats sleeping peacefully.

Edgar was thinking very different thoughts. "Good work!" he whispered to himself with a snicker. "Sweet dreams, my little cats!"

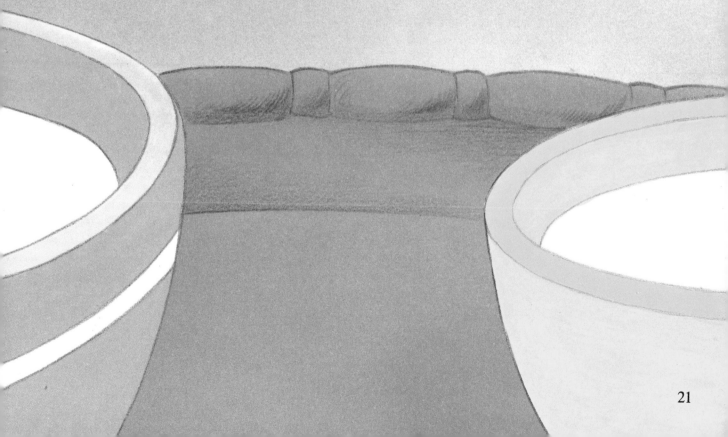

After Madame Bonfamille had gone to bed, Edgar
slipped out of the house with the cats in their basket.
He placed them in the sidecar of his motorcycle and
drove out to the country.

"I will leave them beneath a bridge," he mused.
"They will have fresh air, a roof over their heads, and
water to drink! Country life will be excellent for their
health!"

Everything was working out very well, and Edgar
was so pleased that he did not see Napoleon and
Lafayette, the two guard dogs of a nearby farm.
Napoleon and Lafayette did not like loud motorcycles,
nor did they like the smug look on Edgar's face.

"Let's chase this fellow off the road," suggested
Napoleon.

"Okay, Napoleon. I wouldn't mind some fun," agreed
Lafayette.

"Ready!" barked Napoleon.

"Ready!" growled Lafayette.

The two dogs took off after Edgar on his motorcycle, barking furiously. They took Edgar completely by surprise. He tried to speed away, but the dogs were professional motorcycle chasers. While Lafayette blocked the way, Napoleon snapped at Edgar's ankles.

With a scream, Edgar skidded off the road and zipped down the riverbank by the bridge, with Napoleon and Lafayette right on his tail. He hit a bump and the basket with the cats flew up in the air, landing safely in the reeds by the river.

Edgar's troubles were not over. He sped across the
river and up the other bank, followed by Napoleon and
Lafayette. The two dogs were enjoying the chase too
much to let him go. Snarls, barks, growls, and howls
followed the evil butler down the road.

When they caught up with him, Lafayette took a bite
out of his pants and Napoleon jumped into the sidecar.
Edgar swung at him with his umbrella, but the dog
would not jump out. Edgar had no choice but to kick
the sidecar off the motorcycle. Quickly he sped away
to the safety of the streets of Paris.

Meanwhile, by the riverbank, four frightened pairs of eyes searched the unfamiliar darkness. Duchess and her three kittens had awakened. They stepped out of their basket when the dogs' barking stopped.

"Mom, where are we?" asked Marie, terrified. Duchess did not answer. She did not know.

In Paris, Madame Bonfamille woke up. When she
saw that her cats' basket had disappeared, she searched
the whole house.

"Duchess! Toulouse! Marie! Berlioz! Where are
you?" she called. She rang for Edgar, but he, too, was
nowhere to be found.

Roquefort woke up to Madame's desperate calling for her cats. He was still sleepy, but he crawled out of his hole. Just then Edgar walked in the front door.

Madame ran downstairs when she heard the front door slam. "Edgar! Where were you? Somebody has stolen my children!" cried the distraught old lady.

"Stolen? I don't think so," lied Edgar. "They probably went out for a walk."

"But their basket is gone, too," protested Madame. "I know they were stolen."

Roquefort listened carefully. "Madame is right! Duchess, Berlioz, Toulouse, and Marie have been stolen!" thought the mouse with shock. "Oh! The milk we had for dinner made us sleepy! And Edgar just returned..." Slowly, the little mouse put two and two together. Edgar must have taken the cats away, but why? And where? Roquefort put on his detective cap and set out to get more evidence.

The next morning, Berlioz, Toulouse, and Marie were afraid. "What is going to happen to us?" they cried to their mother. "Where is Madame? We want to go home!"

Duchess did not know what to tell them. All of a sudden, an enormous tomcat jumped out in front of her.

"Hello! My name is Abraham de Lacy Giuseppe Tracy Thomas," he said. "O'Malley for short, and at your service, Ma'am."

O'Malley did not seem very distinguished, but Duchess decided to trust him. As best she could, she explained the strange happenings of the night before.

"I'm just an alley cat, Princess," he said to her with a smile. "But I'll do my best to get you back home!"

"He called me 'Princess'!" thought Duchess. "He knows we are Aristocats. I think he will get us back home."

But Marie was not so sure. "Mother, I don't want to follow him. He looks strange. Where will he take us?" she whimpered.

"He will take us back to Madame," said Duchess firmly. "O'Malley is going to help us get home. Now come on, do as you are told."

The tomcat had spotted a truck by the side of the road. Quietly he ordered them to jump into the back. Just as he leapt up into the truck after them, the engine roared and the truck pulled away, heading for Paris.

But a little farther ahead, the truck came again to a full stop. The driver let out a moan. A breakdown!

"Jump!" ordered O'Malley. If the driver found them in his truck, they would be in trouble.

O'Malley decided to follow the railroad tracks to the city. It was going to be a long walk, and Berlioz, Toulouse, and Marie were not very happy.

"We never should have taken that truck. We should have taken the train!" complained Berlioz.

"This cat doesn't look like the sort to take the train," commented Toulouse.

"Tell him that we are Aristocats!" said Marie, trying to look dignified.

"How ungrateful you are!" scolded Duchess. "You ought to thank him for helping us find our way home!"

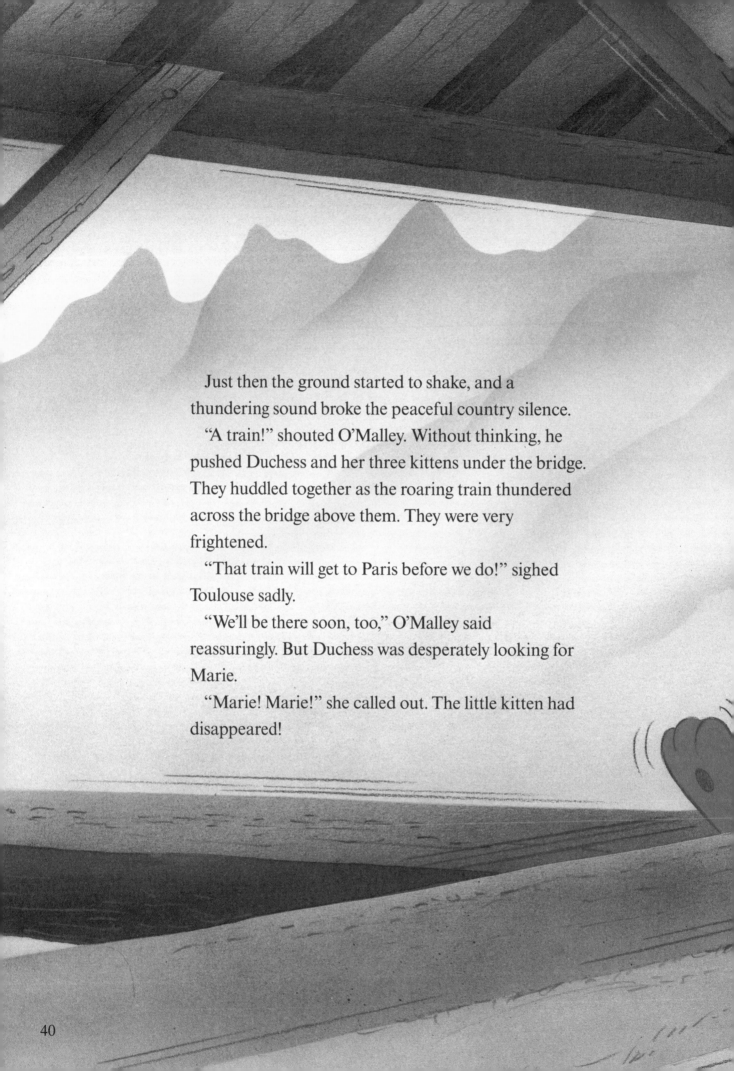

Just then the ground started to shake, and a thundering sound broke the peaceful country silence.

"A train!" shouted O'Malley. Without thinking, he pushed Duchess and her three kittens under the bridge. They huddled together as the roaring train thundered across the bridge above them. They were very frightened.

"That train will get to Paris before we do!" sighed Toulouse sadly.

"We'll be there soon, too," O'Malley said reassuringly. But Duchess was desperately looking for Marie.

"Marie! Marie!" she called out. The little kitten had disappeared!

"Meow! Meow!" Two faint cries were heard below. When the train roared by, Marie had slipped off her perch and fallen into the river.

"Meow! Meow!" she cried as the current carried her away. O'Malley quickly jumped into the water to rescue her.

With a few powerful strokes, he reached Marie and, lifting her above the water, he swam back to shore. Duchess took hold of her daughter.

"O'Malley," she said with tears of gratitude in her eyes, "I cannot thank you enough for this. You have saved my daughter's life."

Marie looked up at the tomcat and shyly whispered a little "Thank you." O'Malley had won her trust.

Two big white geese glided up to O'Malley. They had watched him rescue Marie.

"What a show you put on!" clucked the first.

"An excellent show!" added the second. "I thought cats hated to swim!"

"Who are you?" asked O'Malley, a little flattered.

"We are the Gabble Sisters. I am Emilie and she is Amelie," answered the first. "We are very pleased to meet you, Mr..."

"O'Malley," finished the tomcat. Emilie and Amelie giggled, much to the kittens' amusement.

Once everyone was safely back on the riverbank, the cats began to explain to the Gabble Sisters what had happened to them.

"Paris?" they interrupted, delighted. "We are going to Paris, as well! We've never been to the city! Why don't we all go together?"

Berlioz, Toulouse, and Marie had never seen geese before, and they were fascinated by Emilie and Amelie.

"Oh, yes! Come with us!" they cried out together. "We'd love you to come!" And with that, they all headed towards Paris.

It was close to midnight when they reached the city. O'Malley and the Aristocats said goodbye to the Gabble Sisters.

"Let's take a shortcut home," suggested O'Malley when the geese had left. "Hop on my back, Marie!" he called to the tired kitten. He then led them up to the rooftops, which he used as streets. Marie, Toulouse, and Berlioz had never seen the city from above.

"Wow! This is great!" exclaimed Toulouse.

Meanwhile, Edgar had returned to the countryside to retrieve his sidecar, hat, and umbrella. But Napoleon and Lafayette had made themselves at home with them. They were still laughing about their successful chase.

"This basket is very comfortable," Lafayette told Napoleon. "It's even got cushions! I wonder whose it was."

"Chasing that fellow was worth it," chuckled Napoleon, wearing his new hat. "These are better beds than our old haystack in the barn!"

Edgar sneaked up on them. He had brought his fishing pole to help him get his things back, but there was nothing he could do until the dogs had fallen asleep.

When Napoleon started to snore, Edgar gently lifted him out of the sidecar. He grabbed the hat in his teeth, and put the dog on the haystack. After he nabbed the umbrella, he slipped away. This time he had left his noisy motorcycle far down the road. He did not want to be chased again.

Back in Paris, Roquefort was conducting a thorough investigation. He had spent the day and the whole night interrogating all of the mice in the neighborhood.

Nobody had seen a thing. In the morning, he went to the stables to ask Frou-Frou the mare if she had noticed anything.

"Well, yes, Roquefort, I did notice something strange," said the mare. "The night before last, Edgar left on his motorcycle and came back *without* his sidecar. And last night he left again and came back *with* his sidecar!"

"Thank you, Frou-Frou. I've got a feeling Edgar is up to no good. Let me know if he goes out again," said the mouse.

When Roquefort returned to the house, he was very surprised to see that Duchess and the kittens had returned! They were saying goodbye to their friend when Edgar opened the door. The butler could not believe his eyes. How had they made it back? The cats did not even notice his surprise because they were so happy to be home. But Edgar quickly thought of a way to get rid of them ... forever.

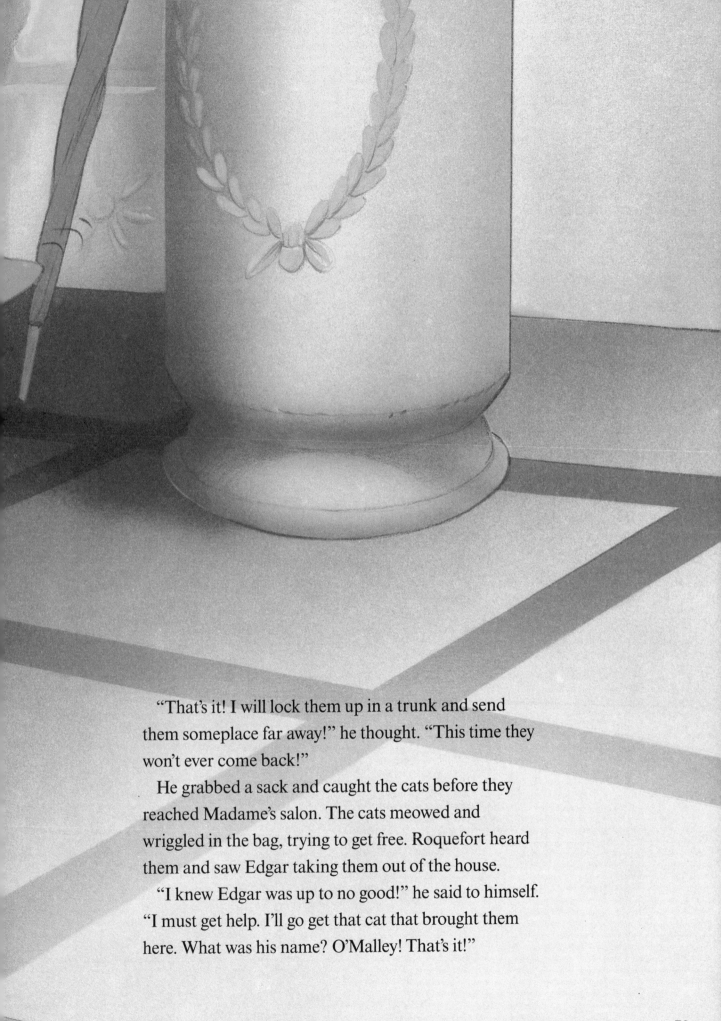

"That's it! I will lock them up in a trunk and send them someplace far away!" he thought. "This time they won't ever come back!"

He grabbed a sack and caught the cats before they reached Madame's salon. The cats meowed and wriggled in the bag, trying to get free. Roquefort heard them and saw Edgar taking them out of the house.

"I knew Edgar was up to no good!" he said to himself. "I must get help. I'll go get that cat that brought them here. What was his name? O'Malley! That's it!"

Roquefort was courageous. He knew that cats did not like mice, especially alley cats, but he had to help his friends. He went after O'Malley. When he caught up with him, he explained what he had seen. O'Malley was furious. He did not like the idea that Duchess and the kittens might get hurt.

"Thanks, Roquefort. You're a brave mouse. I must rescue them before he hurts them, but we'll need help. Go find my friend Scat Cat and his band and tell them to meet me at Madame Bonfamille's house. Quick!" And with that, O'Malley ran back to the house.

Roquefort was a little nervous. Who was Scat Cat? Did he like mice?

When Roquefort found Scat Cat and his band, his heart was pounding furiously. How had the distinguished Duchess and the kittens met these cats? If he didn't quickly explain why he had come, they might just take a bite out of him. Scat Cat picked him up by the tail, while the Siamese cat poked at him.

"You say O'Malley sent you here?" Scat Cat asked the terrified mouse.

"Duchess and the kittens are in great danger," stammered Roquefort. "O'Malley needs you to help save them."

Scat Cat immediately let the mouse go. "Sure we'll help Duchess and her kittens. Show us the way," he ordered. They took off for Madame's house.

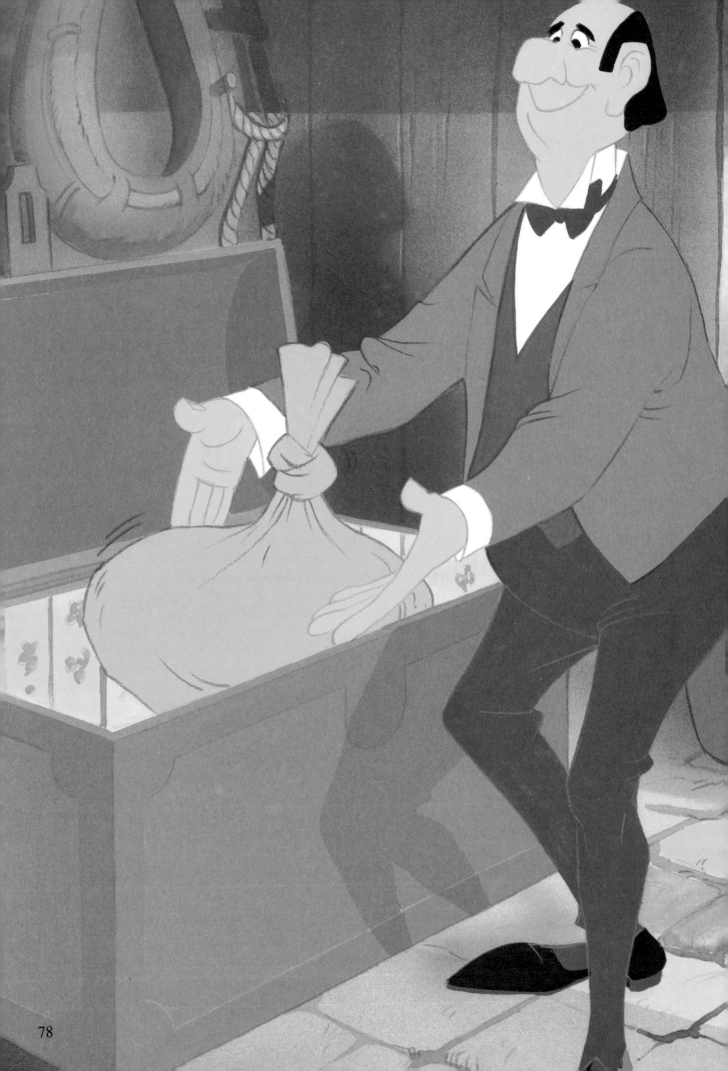

Edgar had taken the cats to the stables. He plopped them in a trunk and congratulated himself on his quick thinking.

"I'm scared!" cried Marie.

"I'm choking in here!" whined Toulouse. Berlioz said nothing, but his heart thumped loudly. Duchess tried to calm them. Just as Edgar closed the trunk, she heard Frou-Frou neighing. "We are in the stables," she realized.

"Frou-Frou! Help!" she cried as the lid closed.

Edgar was sticking a label on the trunk that read "Timbuktu, Africa."

Frou-Frou caught his coat in her mouth. The mare was furious. She pulled and pulled at Edgar's coat while the angry butler struggled to get free.

Frou-Frou wanted to kick him, but she couldn't get close enough. She held on to Edgar as long as she could, neighing for help. O'Malley heard her screams and ran into the stable just as Edgar's coat ripped free.

"That silly mare!" said Edgar. "She thinks she can stop me! I must get this trunk out in time for the pick-up." He pushed it towards the door.

Suddenly O'Malley jumped on top of him, hissing and growling. Edgar fought back, but had no idea who or what he was fighting. All he knew was that he was being scratched again and again. O'Malley fought hard. He would do anything for Duchess and the kittens.

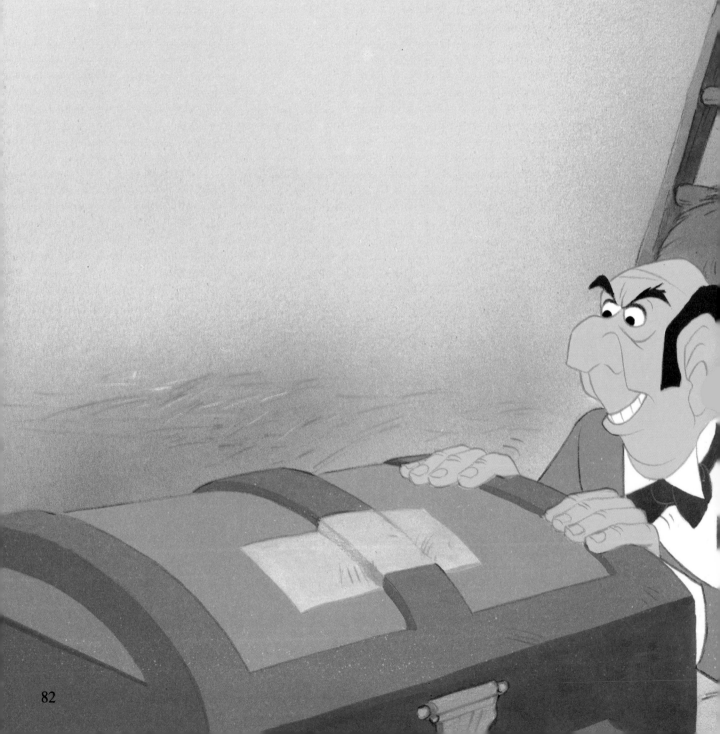

All of a sudden Edgar was back on his feet. He
grabbed a pitchfork and cornered O'Malley.

"I have you now!" the butler hissed. With one swift
jab, he pinned the tomcat to the wall.

When Roquefort tiptoed up the pitchfork's handle and
whispered to the cat, "Scat Cat and his friends are
coming," O'Malley's face lit up. Edgar was not going
to get away now.

The butler sighed. He was not enjoying this.
Everything was getting much too difficult. He
straightened out his clothes and pushed the trunk a
little closer to the door.

"One, two, three, go!" a voice said, and suddenly
what Edgar thought were one hundred cats jumped on
him. They hissed, they bit, they scratched, they
growled, and they pinned him down. Edgar could not
move. Frou-Frou cheered them on. Roquefort held his
breath. Duchess and the kittens were saved!

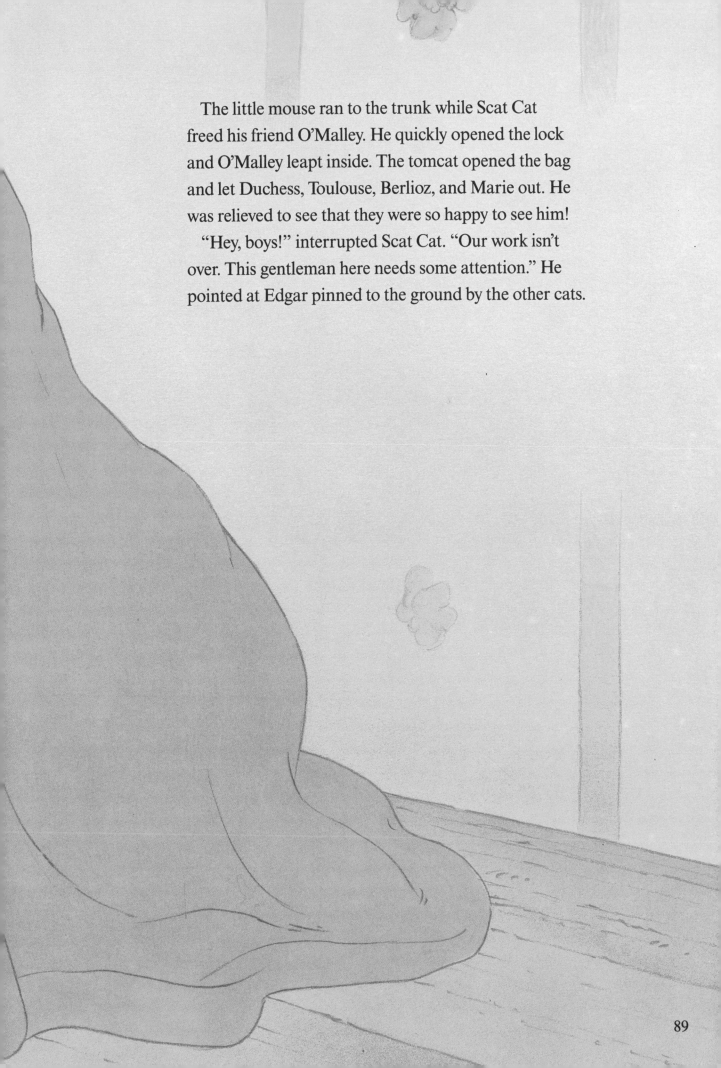

The little mouse ran to the trunk while Scat Cat
freed his friend O'Malley. He quickly opened the lock
and O'Malley leapt inside. The tomcat opened the bag
and let Duchess, Toulouse, Berlioz, and Marie out. He
was relieved to see that they were so happy to see him!

"Hey, boys!" interrupted Scat Cat. "Our work isn't
over. This gentleman here needs some attention." He
pointed at Edgar pinned to the ground by the other cats.

"Why don't we lock *him* up in the trunk and send *him* away?" proposed Frou-Frou. "That way, he'll never come back and bother us."

Everybody thought that Frou-Frou's idea was excellent. Edgar would be sent to Timbuktu! They tied him up, lifted him up with the help of a pulley, and Frou-Frou swiftly kicked him into the trunk. After they had locked it, they pushed the trunk outside to be hauled away by the moving company.

"He'll love those blue skies and palm trees in Timbuktu!" joked Scat Cat when the movers arrived to pick up the trunk.

"Bon voyage!" teased the Siamese cat. Everybody laughed. They happily watched the movers check the label and carry Edgar away.

Duchess and O'Malley said goodbye to Scat Cat and
his band and thanked them for their help. They
promised to come to the attic for another jazz evening
soon.

Madame Bonfamille was very happy when she saw
her cats. Tears and hugs and purrs and furry cuddles
went out for everyone. Madame could see that Duchess
was very fond of O'Malley. She asked him to stay.

With a bit of grooming, O'Malley the alley cat
became Madame's fifth Aristocat. She adopted him and
added him to her will. And when her five children
posed for a photograph, what a handsome family they made!

ISBN 1-57082-032-5
10 9 8 7 6 5 4 3 2 1